Time Series Analysis - New Insights

Edited by Rifaat Abdalla,
Mohammed El-Diasty, Andrey Kostogryzov
and Nikolay Makhutov

Published in London, United Kingdom

Time Series Analysis - New Insights
http://dx.doi.org/10.5772/intechopen.100724
Edited by Rifaat Abdalla, Mohammed El-Diasty, Andrey Kostogryzov and Nikolay Makhutov

Contributors
Andrey Kostogryzov, Nikolay Makhutov, Andrey Nistratov, Georgy Reznikov, Tom Burr, Kim Kaufeld, Monday Osagie Adenomon, Felicia Oshuwalle Madu, Igor N. Sinitsyn, Anatoly S. Shalamov, Abdelouahab Bibi, Sameera Othman, Haithem Mohammed Ali, Guy Mélard, Rifaat Abdalla, Mohammed El-Diasty

First published in London, United Kingdom, 2023 by IntechOpen
IntechOpen is the global imprint of INTECHOPEN LIMITED, registered in England and Wales, registration number: 11086078, 5 Princes Gate Court, London, SW7 2QJ, United Kingdom

British Library Cataloguing-in-Publication Data
A catalogue record for this book is available from the British Library

Additional hard and PDF copies can be obtained from orders@intechopen.com

Time Series Analysis - New Insights
Edited by Rifaat Abdalla, Mohammed El-Diasty, Andrey Kostogryzov and Nikolay Makhutov
p. cm.
Print ISBN 978-1-80356-305-3
Online ISBN 978-1-80356-306-0
eBook (PDF) ISBN 978-1-80356-307-7

For EU product safety concerns:
IN TECH d.o.o., Prolaz Marije Krucifikse Kozulić 3, 51000 Rijeka, Croatia,
info@intechopen.com or visit our website at intechopen.com.

Meet the editors

Dr. Rifaat Abdalla has been an associate professor in the Department of Earth Sciences, College of Science, Sultan Qaboos University, since 2017. He specializes in geoinformatics, with a focus on WebGIS applications and remote sensing modeling. For the previous five years, he had been at King Abdulaziz University, Jeddah. He has also worked in the oil industry in Qatar and in the UK, and has held teaching posts at York University and Ryerson University, Canada. Dr. Abdalla has GIS Professional (GISP) and Professional Geoscientist (P.Geo.) certifications from Ontario, Canada. His research interests include hydrography and marine applications for GIS; modeling and simulation; mobile handheld GIS visualization; and disaster management and emergency response mapping. Dr. Abdalla is a recipient of several prestigious international awards, including the American Society for Photogrammetry and Remote Sensing (ASPRS) PE&RS Best Scientific Paper by ESRI.

Mohammed El-Diasty has been an associate professor in the Department of Civil and Architectural Engineering, College of Engineering, Sultan Qaboos University, since 2021. He was previously at Mansoura University, Egypt for twelve years and at King Abdulaziz University, Saudi Arabia for nine years. He received his MSc degree in civil engineering from Ryerson University, Canada and his Ph.D. in geomatics engineering from York University, Canada. Dr. El-Diasty's Certified Professional Engineer (P.Eng.) status was awarded in Ontario, Canada. Dr. El-Diasty specializes in geomatics engineering. His research interests include positioning, navigation, mobile mapping, GNSS/INS integration, hydrography and gravimetry. He has received several research grants and has published scholarly scientific manuscripts in geomatics engineering research and practical fields.

Andrey Kostogryzov is the Chief Researcher at the Computer Science and Control Federal Research Center of the Russian Academy of Sciences, Moscow. He obtained his doctorate of engineering in 1994 and was appointed a professor in 1999. He is deputy chairman of the Committee on Business Safety and chairman of the Information Technologies Subcommittee of the Russian Federation Chamber of Commerce and Industry. He is also chairman of the Special Expert Council of the Higher Attestation Commission, and a Certified Expert of the Russian Academy of Sciences, the Ministry of Education, and Gazprom. Dr. Kostogryzov was named an Honored Science Worker of the Russian Federation and received a Laureate of the Russian Government Award in the field of science and engineering in 2015. He is the author of more than 100 probabilistic models and more than 200 scientific works, including 21 books.

Nikolay Makhutov is the Chief Researcher at the A.A. Blagonravov Institute for Machine Science of the Russian Academy of Sciences. He obtained his doctorate in technical sciences in 1974 and was appointed a professor in 1978. He has been a corresponding member of the Russian Academy of Sciences since 1987, is a Laureate of the State and has been awarded three Russian Federation government prizes in the field of science and technology. He specializes in the strength and safety of complex technical systems (nuclear, thermal and hydraulic power, rocket-space and transport complexes). He is the author of more than 600 scientific publications (journal articles, reports) and 40 monographs published in Russia and abroad, and a member of the editorial boards of Russian and foreign scientific books and journals.

Contents

Preface

A time series is simply a sequence of data points occurring in succession for a given period of time. Time series data are a collection of observations obtained through repeated measurements over time. Time series data are of two types: measurements gathered at regular time intervals (metrics); and measurements gathered at irregular time intervals (events).

Time series data are everywhere since time is a constituent of everything that is observable. As our world becomes increasingly digitized, sensors and systems are constantly emitting a relentless stream of time series data which have numerous applications across various industries. Graphs of time series data points can often illustrate trends or patterns in a more accessible, intuitive way.

Time series analysis is a specific way of analyzing a sequence of data points that are collected and recorded at consistent intervals over a set period of time rather than just recorded intermittently or randomly. Time series analysis typically requires a large number of data points to ensure consistency and reliability. It also ensures that any trends or patterns discovered are not outliers and can account for seasonal variance. Additionally, time series data can be used for forecasting and predicting future data based on historical data. Time series analysis is used for non-stationary data, things that are constantly fluctuating over time or are affected by time. Industries like finance, retail, and economics frequently use time series analysis. New time series analysis tools are needed in disciplines as diverse as astronomy, economics, and meteorology. Examples of time series analysis presented in this book include:

- Digital elevation model error analysis and modeling combining principal component analysis and least-squares method
- ARIMA models with time-dependent coefficients
- Methods of conditionally optimal forecasting for stochastic continuous acquisition logic support technologies
- A two-stage modeling framework for time series analysis of spatiotemporal data
- Forecasting functional nonparametric time series using the parametric power transformation
- Monitoring residuals from time series models
- Comparison of the out-of-sample forecast for inflation rates using ARIMA and ARIMAX models
- The subordinated solution of continuous-time bilinear time series.

Among other applications of time series analysis are quarterly sales, weather forecasting, rainfall measurement, heart rate monitoring (EKG) and brain monitoring (EEG).

Because time series analysis includes many categories or variations of data, analysts must sometimes make complex models. However, not all variances can be accounted for, and models that are too complex or that try to do too many things can lead to a lack of fit. Lack of fit or overfitting models may fail to distinguish between random error and true relationships, leaving analysis skewed and forecasts incorrect.

Time series analysis models include the following types:

- *Classification:* Identifies and assigns categories to the data.
- *Curve fitting*: Plots the data along a curve to study the relationships of variables within the data.
- *Descriptive analysis*: Identifies patterns in time series data, like trends, cycles, or seasonal variation.
- *Explanative analysis*: Attempts to understand the data and the relationships within it, as well as cause and effect.
- *Exploratory analysis*: Highlights the main characteristics of the time series data, usually in a visual format.
- *Forecasting*: Predicts future data. This type is based on historical trends. It uses historical data as a model for future data, predicting scenarios that could happen along future plot points.
- *Intervention analysis*: Studies how an event can change the data.
- *Segmentation*: Splits the data into segments to show the underlying properties of the source information.

There are a few factors that can cause variations in time series data. The following five components are used to describe how time series data behaves:

- *Autocorrelation* refers to the relationship between a given observation in a time series and a previous observation in the same time series, where the interval between the two data points is referred to as a "lag." Analysts use autocorrelation functions to understand if the connection between two lags is significant and to determine how random or stationary the time series is.
- *Seasonality* is when data experiences predictable changes at regular intervals such as quarterly, monthly, or biannually. For instance, summer clothes are sold more in the spring than in other seasons, and Black Friday is the busiest shopping day of the holiday season. Seasonality always occurs in a fixed and known period.

- A *trend* represents a long-term movement of data in a certain direction. The trend can be increasing or decreasing or even linear or nonlinear. Not all series have a noticeable trend—things like fires, floods, revolutions, earthquakes, strikes, and epidemics are clear representations of this. That said, the overall trend must be upward, downward, or stable. Examples include periods of economic growth and recession, the average prices of apartment rentals in each city, and sales of a particular product.

- *Cycles* occur when data show a rise-and-fall pattern that is not over a fixed period. Many people confuse cyclical variations with seasonal variations, but they are quite different. Cyclical variations have nothing to do with the time of year and cannot be measured according to a given calendar month. They also typically last longer than seasonal variations and are often economic in nature. For example, monthly housing sales can reflect overall market trends, and demand rises and falls in a cyclical pattern over time.

- An *irregular* component is due to short-lived fluctuations in a series. While they are not predictable, sometimes irregularities such as sales tax changes can be anticipated. Irregularities represent the remaining time series outside the trend cycle and the seasonal components. Examples include natural disasters, health crises, and wars.

Rifaat Abdalla
Department of Earth Sciences,
College of Science,
Sultan Qaboos University,
Al-Khoudh, Oman

Mohammed El-Diasty
Sultan Qaboos University,
Al-Khoudh, Oman

Andrey Kostogryzov and Nikolay Makhutov
Russian Academy of Sciences,
Moscow, Russia

Section 1

Time Series Data Modeling

Chapter 1

Sensitivity Analysis and Modeling for DEM Errors

Mohammed El-Diasty and Rifaat Abdalla

Abstract

The Digital Elevation Model (DEM) can be created using airborne Light Detection And Ranging (LIDAR), Image or Synthetic-Aperture Radar (SAR) mapping techniques. The direct georeferencing of the DEM model is conducted using a GPS/inertial navigation system. The airborne mapping system datasets are processed to create a DEM model. To develop an accurate DEM model, all errors should be considered in the processing step. In this research, the errors associated with DEM models are investigated and modeled using Principal Component Analysis (PCA) and the least squares method. The sensitivity analysis of the DEM errors is investigated using PCA to define the significant GPS/inertial navigation data components that are strongly correlated with DEM errors. Then, the least squares method is employed to create a functional relationship between the DEM errors and the significant GPS/inertial navigation data components. The DEM model errors associated with airborne mapping system datasets are investigated in this research. The results show that the combined PCA analysis and least squares method can be used as a powerful tool to compensate the DEM error due to the GPS/inertial navigation data with about 27% in average for DEM errors produced by the direct georeferenced airborne mapping system.

Keywords: sensitivity, PCA, least squares, DTM errors, navigation

1. Introduction

The Digital Elevation Model (DEM) can be created using airborne Light Detection And Ranging (LIDAR), Image or Synthetic-Aperture Radar (SAR) mapping techniques. The direct georeferencing of DEM model is conducted using GPS/inertial navigation system. The accuracy of the developed DEM model is strongly dependent on the mapping system and the georeferencing system grades. The selection of the mapping system and the georeferencing system grades is carried out in the planning stage based on the accuracy requirements of the required DEM model.

The previous literatures focused and heavily investigated the DEM model generation based on LIDAR, Images, and SAR data cleaning and filtering techniques such as Triangulated Irregular Network (TIN)-based filtering, slope-based filtering, mathematical morphological filtering, interpolation-based filtering, and Machine-learning-based filtering. The original TIN-based filtering was developed based on the classical progressive TIN densification (PTD) and was implemented effectively in the commercial software TerraScan [1, 2]. Then, the revised PTD was investigated and reduced the total errors by about 8% when compared with classical PTD method [3]. Afterwards, a Parameter-Free PTD (PFPTD) algorithm was developed and outperforms the classical and revised PTD methods [4]. The original slope-based filtering was derived based on height differences in the training dataset [5, 6]. Then, adaptive slope-based filtering algorithm was developed to improve the accuracy in urban applications when compared with the original slope-based filtering algorithm [7]. The original mathematical morphological filtering was proposed to filter LIDAR data [8]. Then, the progressive morphological filtering algorithm was developed to improve the original method by applying threshold condition based on the elevation differences and proposed increasing gradually the filtering window size [9]. Afterward, the spline iteration method was introduced to improve the morphological filtering algorithm [10]. All mathematical morphological filtering methods outputs are strongly dependent on adapting the filtering window size. The original interpolation-based filtering method was proposed to deal with the steep areas [11]. Then, the interpolation-based DEM generation method was developed where one of the Inverse Distance Weighted, Kriging, and Natural Neighbor (NN) can be employed for DEM generation [12]. The Natural Neighbor (NN) method was proven to provide most efficient results. Finally, Machine-learning-based filtering was investigated where this method depends on topographic characteristics of the areas under investigation [13]. The deep convolutional neural network (CNN) was proposed for develop accurate DEM model [14]. The Machine learning method was optimized using windowing method to improve the DEM model generation [15].

However, the above-mentioned methods are mainly dependent on the data cleaning and filtering techniques of the heights to develop the DEM model. These methods are associated with errors that are not considered in the DEM modeling and could be corelated with system navigation data. Therefore, this research investigated the sensitivity analysis modeling of DEM errors that are potentially correlated with navigation data to improve the overall accuracy of the DEM model.

2. Combined PCA and least squares method

The sensitivity and modeling of DEM errors are investigated in this research using PCA and least squares function modeling. The system navigation data (position, velocities, attitudes, accelerations and dopplers) are considered the inputs to the model and the DEM height error is considered the desired output of the model.

PCA is a numerical technique used to study multidimensional processes that can be used to (1) reduce the dimensionality of a dataset and (2) identify relationships between the underlying variables of the process. PCA is based on eigen or singular value analysis of the process correlation or covariance matrix. The goal of PCA is to determine the minimum number of eigenvectors that best describe the key features of the process correlation matrix. This results in a reduced-dimensionality model for the matrix which can be used for data analysis, reduction, and model synthesis. Singular value decomposition (SVD) is fundamental to PCA. More details on PCA and SVD can also be found in Jolliffe [16].

DOI: http://dx.doi.org/10.5772/intechopen.108788

2.1 PCA analysis

Let X denote an m x n matrix. For convenience, we assume $m \geq n$. The elements of the ith row of X form the n-dimensional vector $\mathbf{g}_i$. The elements of the jth column of X form the m-dimensional vector $\mathbf{a}_j$. The general singular value decomposition (gSVD) of X can be written as:

$$X = U\Sigma V^T \tag{1}$$

where U is an m x m matrix, Σ is an m x n matrix containing the singular values, and V^T is an n x n matrix. The columns of U are called the *left singular vectors*, $\{\mathbf{u}_k\}$, and form an orthonormal basis for the range space, so that $\mathbf{u}_i \cdot \mathbf{u}_j = 1$ for $i = j$, and $\mathbf{u}_i \cdot \mathbf{u}_j = 0$ otherwise. The rows of V^T contain the elements of the *right singular vectors*, $\{\mathbf{v}_k\}$, and form an orthonormal. U and V are orthonormal so that their inverses exist and are their transposes. The matrix Σ can be decomposed as:

$$\Sigma = \begin{bmatrix} S \\ 0 \end{bmatrix} \tag{2}$$

where S is an m x n diagonal matrix in which only the diagonal elements are non-zero, $S = \text{diag}(s_1,\ldots,s_n)$ where the diagonal elements are zero. If the rank of X is r, $s_k > 0$ for $1 \leq k \leq r$, and $s_i = 0$ for $(r + 1) \leq k \leq n$. [For problems like the one we are interested in, noise generally ensures that $r = n$.] By convention, the ordering of the singular vectors is determined by high-to-low sorting of singular values, with the highest singular value in the upper left index of the S matrix. Note that for a square, symmetric matrix X, SVD is equivalent to eigenvalue decomposition. In PCA, the right singular vectors are frequently called the *components*. While the scaled left singular vectors $\{s_i\mathbf{u}_k\}$ are called the *scores*.

Note that U can be decomposed into two submatrices, an m x n matrix U_R and an m x m-n matrix U_N where $U = [U_R\ U_N]$. U_R defines the range space of U, while U_N defines the null space. Note that $X = U_R S V^T$ so that $XV = U_R S$. This provides the reduced form of SVD often used in PCA. In practice this is the form generally used; hence, we often drop the R subscript on U. **Figure 1** illustrates the various reduced-form matrices. Note that the right singular vectors span the space of the row vectors $\{\mathbf{g}_i\}$ and the left singular vectors span the space of the column vectors $\{\mathbf{a}_j\}$.

Several relationships can be derived. The SVD equation for $\mathbf{g}_i$ is:

$$\boldsymbol{g_i = \sum_{k=1}^{r} u_{ik} s_k v_k} \tag{3}$$

which is a linear combination of the right singular values $\{\mathbf{v}_k\}$. The ith row of U, $\mathbf{g'}_i$, contains the coordinates of the ith entry in the coordinate system (basis) of the scaled right singular values, $s_k\mathbf{v}_k$. If $r < n$ (or if we truncate the singular values to $r = l$), this computation requires fewer variables using $\mathbf{g'}_i$ rather than $\mathbf{g}_i$, thus reducing the dimension of the problem. Similarly, the SVD equation for $\mathbf{a}_j$ (the jth column of X) is:

$$\boldsymbol{a_j = \sum_{k=1}^{r} v_{jk} s_k u_k} \tag{4}$$

which is a linear combination of the left singular values $\{\mathbf{u}_k\}$. The jth column of V^T, $\mathbf{a'}_j$ (see **Figure 1**), contains the coordinates of the jth column of X in the coordinate

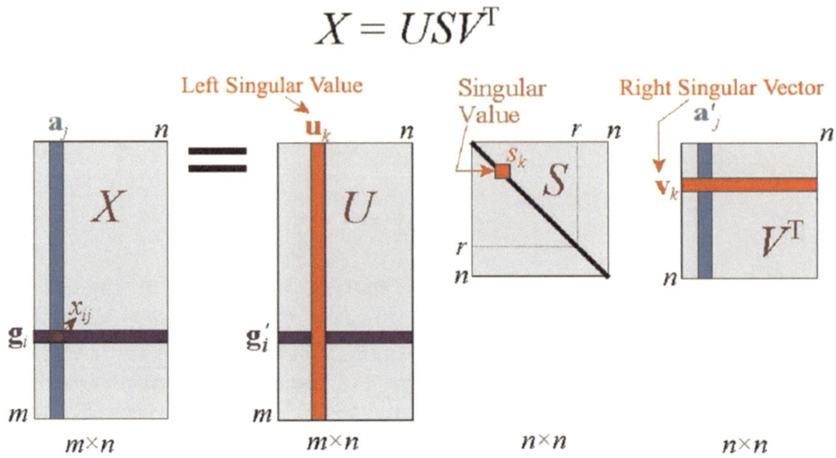

Figure 1.
Illustration of reduced SVD matrices. The right singular vectors have often termed the components while the left singular values are the scaled scores.

system (basis) of the scaled left singular vectors (the scores), $s_k\mathbf{u}_k$. By using the vector $\mathbf{a}'_j$, the analysis may be captured by $r \leq n$ variables, which is always fewer than the m elements in the vector $\mathbf{a}_j$, thus SVD reduces the number of variables required. Essentially, there are only r (which we can truncate to eliminate small singular values and further reduce the dimensionality) component vectors (the corresponding right singular vectors) which explain the behavior of X. The application of PCA often use the SVD property:

$$\boldsymbol{X}^{(l)} = \sum_{k=1}^{l} \boldsymbol{u}_k s_k \boldsymbol{v}_k^T \tag{5}$$

where $X^{(l)}$ is the closest rank-l matrix to X, i.e., $X^{(l)}$ minimizes the sum of the squares of the difference between the elements of X and $X^{(l)}$, diff = $\sum_{ij}|x_{ij} - x^{(l)}{}_{ij}|^2$.

We can define the *covariance matrix* as $X^TX = \Sigma_i \boldsymbol{g}_i \boldsymbol{g}_i{}^T$. SVD analysis of X^TX yields V^T, which contains the principal components of $\{\mathbf{g}_i\}$, i.e. the right singular vectors $\{\mathbf{v}_k\}$ are the same as the principal components of $\{\mathbf{g}_i\}$. The eigenvalues of X^TX are equivalent to $s_k{}^2$, which are proportional to the variances of the principal components. The matrix $XX^T = \Sigma_j \boldsymbol{a}_j \boldsymbol{a}_j{}^T$ is proportional to the covariance matrix of the variables of $\mathbf{a}_j$. The left singular vectors $\{\mathbf{u}_k\}$ are the same as the principal components of $\{\mathbf{a}_j\}$. The $s_k{}^2$ are proportional to the variances of the principal components. The diagonal values of S (*i.e.*, s_k) are the "singular value spectrum". The value of a singular value is indicative of its importance in explaining the data. More specifically, the square of each singular value is proportional to the variance explained by each singular vector.

In PCA, X is defined and the SVD is computed. The singular values s_k are then plotted versus k. A reduced dimensionality approximation to X is computed by truncating the singular value series, i.e. by setting $s_k = 0$ for $k > K$ where K is the chosen threshold value, and using Eq. (3). Note that the squared singular values $s^2{}_k$ are a measure of the variability or "power" in the corresponding signal component specified by the corresponding to singular vector (like the frequency component in Fourier analysis).

2.2 Least squares functional modeling with PCA optimization

We ultimately seek a model that relates the inputs (navigation parameters) to the height error. The least squares method is used to model DEM errors. More details on the least squares method can also be found in Ghilani [17]. Assuming a linear forward model:

$$Xh = e \tag{6}$$

The set of navigation data parameters is used to form the h vector. We will assume that the corresponding DEM height errors form the e vector. For each epoch, we form a single h and e vector. Since the model parameters vary with time, vectors are created by stacking the values as a function of time.

Using the matrix form, the $\boldsymbol{X}$ matrix in Eq. (6) represents the mapping between the model parameters and the height. This mapping is what we want to estimate from the DEM error training set. While there are several approaches, the following approach is attractive. Since Eq. (6) must apply for all realizations we create matrices $\boldsymbol{H}$ and $\boldsymbol{E}$ by combination of all the vectors of h and e, i.e., $\boldsymbol{H} = [\boldsymbol{h}_1|\boldsymbol{h}_2| \ldots \boldsymbol{h}_n]$ and similarly for $\mathbf{E}$. We can then write:

$$\boldsymbol{XH} = \boldsymbol{E} \tag{7}$$

Shifting the mean and scaling the vector elements affects the performance of the estimate. Assuming we have enough realizations, we can write a least-squares empirical estimate of $\boldsymbol{X}$, $\boldsymbol{X}_\mathbf{e}$, as:

$$\boldsymbol{X}_e = \boldsymbol{E}\boldsymbol{H}^T\left(\boldsymbol{H}\boldsymbol{H}^T\right)^{-1} \tag{8}$$

While we can use $\boldsymbol{X}_\mathrm{e}$ directly, SVD analysis of $\boldsymbol{X}_\mathrm{e}$ provides us with a powerful tool to understand the relationship between $\mathbf{H}$ and $\mathbf{E}$. The SVD analysis of $\boldsymbol{X}$ is:

$$\boldsymbol{X}_e = \boldsymbol{U\Sigma V}^T \tag{9}$$

The singular value spectrum of $\boldsymbol{\Sigma}$ tells us the dimensionality (say l) of the model parameter vector (h) are "useful" while the first l column vectors of $\boldsymbol{V}$ provide a basis for the space of "useful" subspace of model parameters. (In the ideal case, a retained column vector of $\boldsymbol{V}$ that contained a single 1 in a particular location w and zero elsewhere would suggest that only the wth component of model vector needs to be used. In general, the l useful columns of $\boldsymbol{V}$ provide a mapping (rotation and scaling) from the original model parameter space to a restricted model parameter space. As noted previously, the value of l is either the rank or is selected by truncating the singular value spectrum. The first l columns of $\boldsymbol{U}$ form a basis of the output or the range space of $\boldsymbol{X}_\mathbf{e}$, that is, of the height error function. The columns of $\boldsymbol{U}$ and $\boldsymbol{V}$ greater than l can be discarded, which simplifies usage of the truncated model.

Note that while the order of the model parameters or error values within the vectors is not important, the range of cases included does matter. The model may be unable to adequately represent a case not included in the training set. Also, the model can only represent linear functions of the input model parameters. (Recall, that we can produce non-linear responses by including non-linear transforms of model parameters in the model parameter vector.)

Let X_r be the singular value reduced estimate of X_e, i.e. Eq. (9) where the diagonal elements of $\boldsymbol{\Sigma}$ are set to zero beyond the lth element. To apply a height correction, we form an h vector of the model parameters from a new data take, and compute an estimate of the height error vector **e**:

$$X_f h = e \tag{10}$$

The error e is subtracted from the height map to remove the height error. Note that due to our formulation of the original model and height vectors that this is done "block-wise".

In this research, the sensitivity analysis is carried out using PCA and the modeling is investigated the least squares function modeling where the system navigation data are considered the inputs to the model and the DEM height error is considered the desired output of the model.

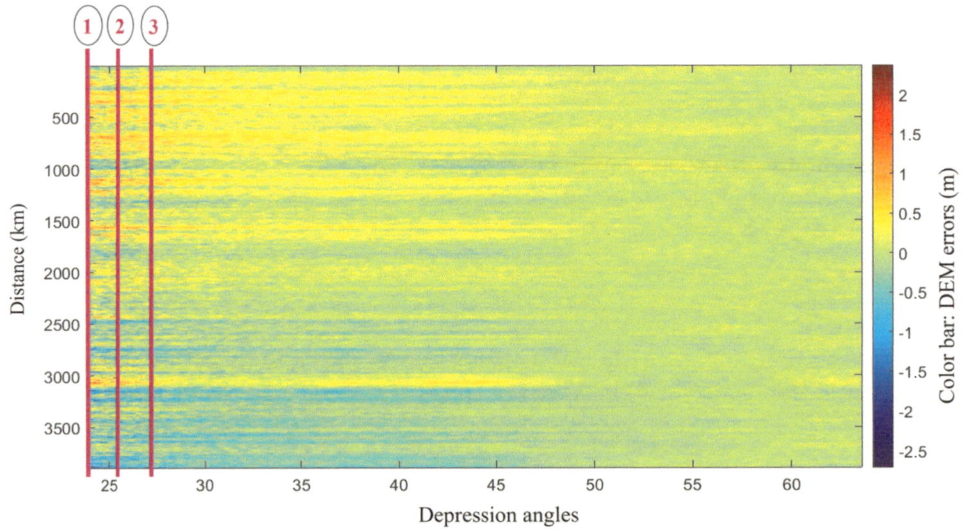

Figure 2.
DEM error map along with the locations of the across-track sections in purple solid lines (1, 2, and 3).

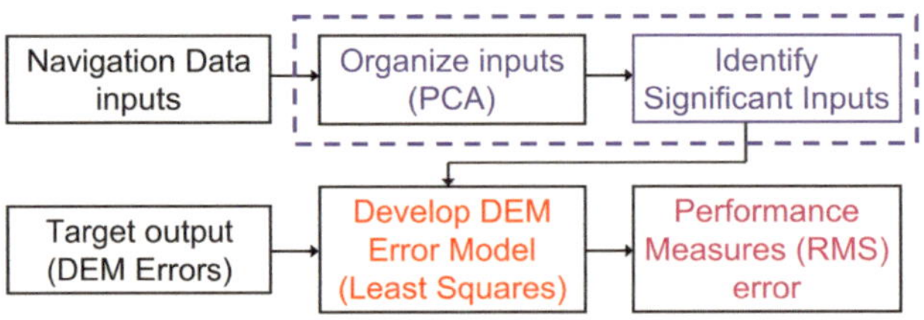

Figure 3.
Combined PCA analysis and least squares methodology.

3. Data and methodology

The DEM model errors associated with airborne mapping system datasets are investigated in this research. **Figure 2** shows an example of DEM errors case study. It can be seen that the DEM errors map shows large errors in the left side associated with low depression angles and small errors in the right side associated with high depression angles. The across-track DEM errors sections at three different depression angles are investigated (sections 1 to 3) as shown in **Figure 2** where the DEM errors are the highest in these locations.

The methodology for the proposed combined PCA analysis and least squares method is shown in **Figure 3**. The PCA analysis is utilized to identify the significant inputs from multiple navigation data and the least squares method is implemented to estimate the DEM errors models. The root mean squares (RMS) errors are used to quantify the accuracy of the developed DEM errors models.

4. Results and discussion

Three across-track sections (1 to 3) as shown in **Figure 2** have been investigated to test the performance of combined PCA analysis and least squares method. The navigation data (3 positions, 3 velocities, 3 attitudes, 3 accelerations, 3 attitude rates, 3 attitudes accelerations, 1 doppler, and 1 doppler rate) represent the input parameters for DEM errors modeling were investigated using PCA analysis method. Out of 20 inputs, 10 inputs were found significant (2 positions, 2 accelerations, 2 attitudes, 2 attitude rates, 1 doppler and 1 doppler rate) in the three across-track sections.

Then the least squares were employed to model the functional relationship between the 10 significant navigation inputs and the targeted DEM error output. The RMS errors were estimated before and after modeling to test the performance of the developed model to compensate the DEM errors. **Figure 4** shows the targeted DEM errors and the modeled DEM errors where **Figure 5** shows the differences between the

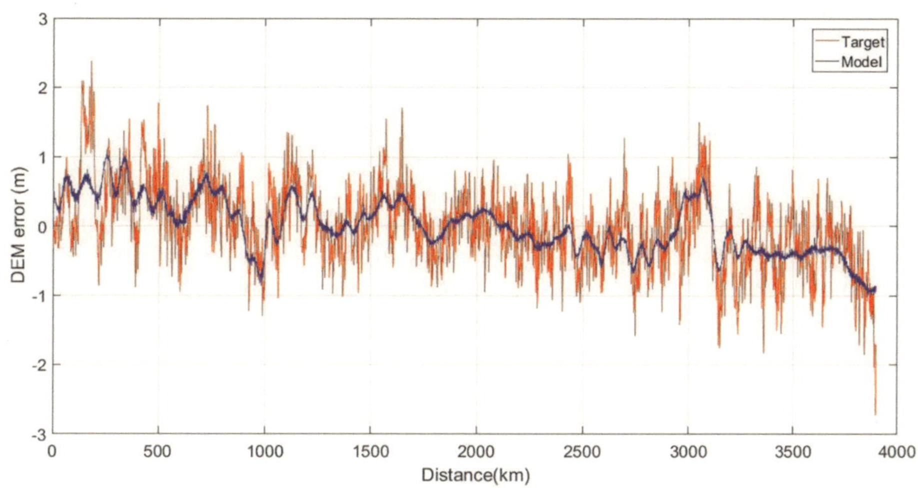

Figure 4.
Section (1) DEM errors where blue line represents the targeted DEM error and red lines represents the modeled DEM error.

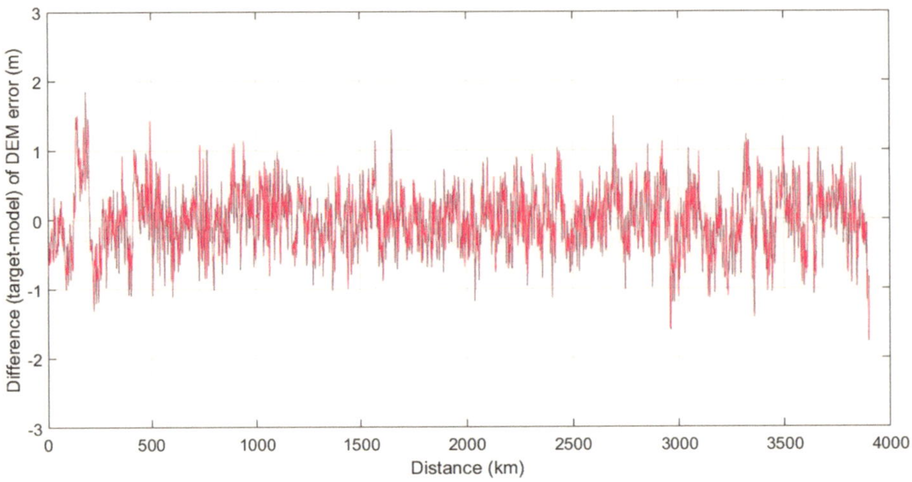

Figure 5.
Section (1) difference between the targeted DEM error and modeled DEM errors.

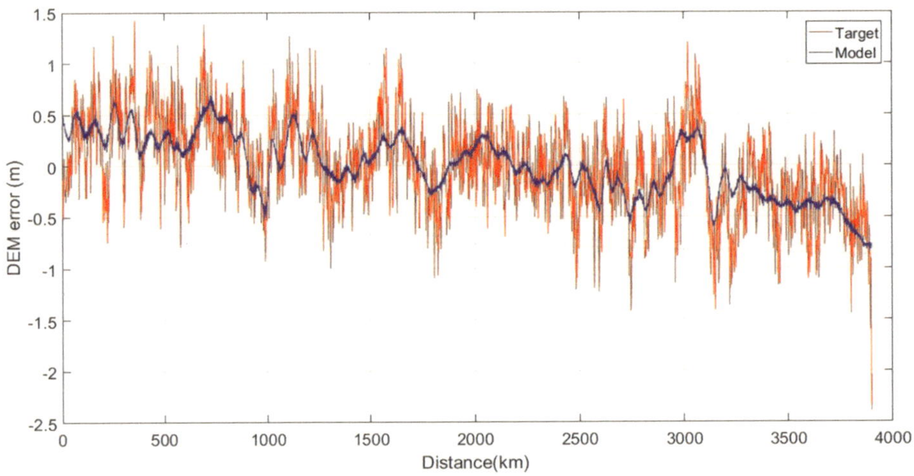

Figure 6.
Section (2) DEM errors where blue line represents the targeted DEM error and red lines represents the modeled DEM error.

targeted and modeled DEM errors for section (1) case. In section (1) case, the estimated RMS error before modeling is 0.59 m and after modeling is 0.44 m which means that about 25% of DEM errors can be compensated in section (1) case. **Figure 6** shows the targeted DEM errors and the modeled DEM errors where **Figure 7** shows the differences between the targeted and modeled DEM errors for section (2) case. In section (2) case, the estimated RMS error before modeling is 0.45 m and after modeling is 0.33 m which means that about 27% of DEM errors can be compensated in section (2) case. **Figure 8** shows the targeted DEM errors and the modeled DEM errors where **Figure 9** shows the differences between the targeted and modeled DEM errors for section (3) case. In section (3) case, the estimated RMS error before modeling is 0.35 m and after modeling is 0.25 m which means that about 28% of DEM errors can be compensated in section (3) case.

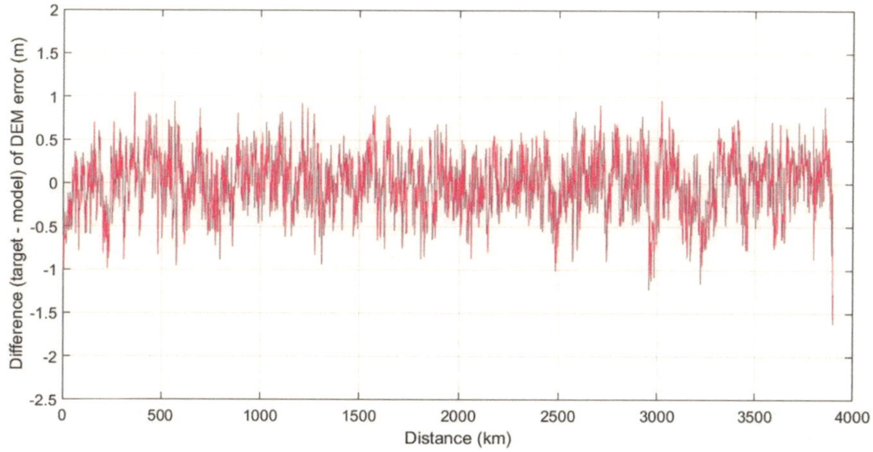

Figure 7.
Section (2) difference between the targeted DEM error and modeled DEM errors.

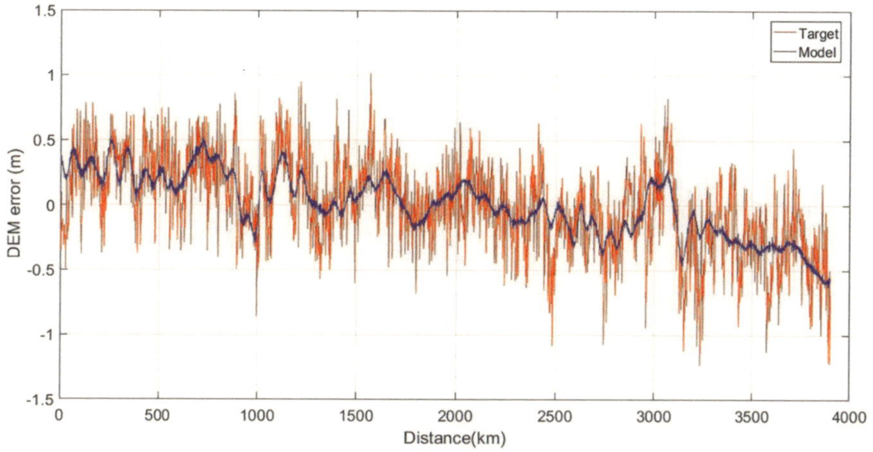

Figure 8.
Section (3) DEM errors where blue line represents the targeted DEM error and red lines represents the modeled DEM error.

Section	RMS error before modeling (m)	RMS error after modeling	Percentage of DEM error compensation (%)
Section (1)	0.59	0.44	25
Section (2)	0.45	0.33	27
Section (3)	0.35	0.25	28
Overall percentage of DEM error compensation			27

Table 1.
RMS errors for all sections before and after DEM error modeling.

Table 1 and **Figure 10** summarize the RMS errors and the percentage of DEM error compensation from the three sections where the overall percentage of DEM error compensation is about 27% on average. The results show that the combined PCA

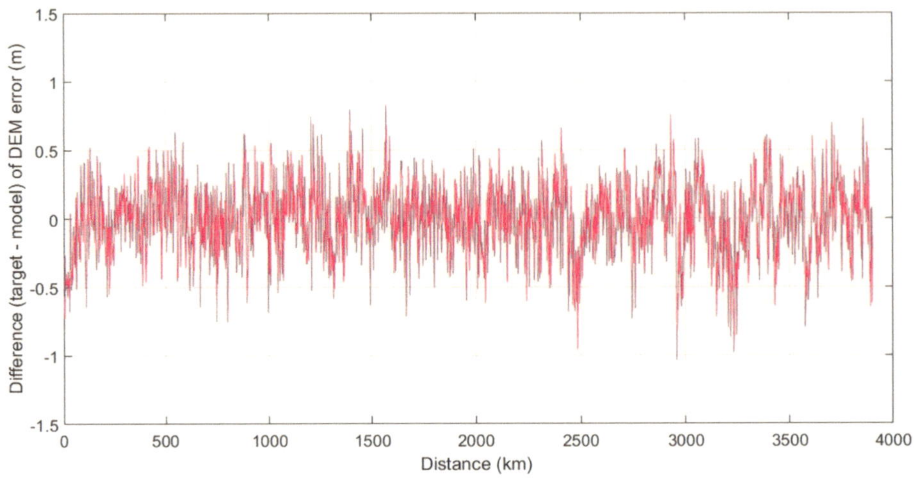

Figure 9.
Section (3) difference between the targeted DEM error and modeled DEM errors.

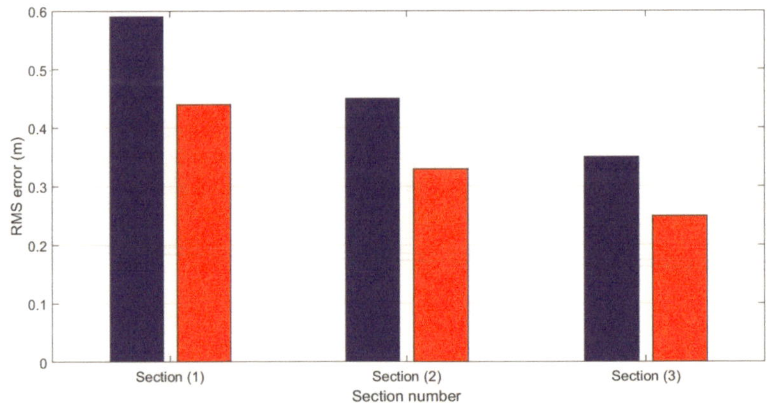

Figure 10.
RMS errors for all sections before and after DEM error modeling.

analysis and least squares method can be used as a powerful tool to compensate the DEM error produced by the navigation data with about 27% in average for the case study investigated in this research.

5. Conclusions and recommendation

The sensitivity analysis of DEM errors to the GPS/inertial navigation data was investigated in this research. It was concluded that the sensitivity analysis of the DEM errors can be performed using the PCA to identify the significant GPS/inertial navigation data components that are strongly correlated with DEM errors. Also, it is concluded that the least squares method can be rigorously utilized to establish the functional relationship between the DEM errors and the significant GPS/inertial navigation data components. The combined PCA and least squares method were validated using the DEM model errors associated with airborne mapping system datasets using three across-track sections of the DEM errors datasets. The results show that the combined PCA analysis and least squares method can be used as a powerful tool to compensate for the DEM error produced by the GPS/inertial navigation data with about 27% on average. Therefore, it is recommended to use the combined PCA analysis and least squares method to reduce the DEM errors associated with the DEM model produced by the direct georeferenced airborne mapping system.

Conflict of interest

The author declares no conflict of interest.

Author details

Mohammed El-Diasty[1*] and Rifaat Abdalla[2]

1 Civil and Architectural Engineering Department, College of Engineering, Sultan Qaboos University, Muscat, Oman

2 Department of Earth Sciences, College of Science, Sultan Qaboos University, Muscat, Oman

*Address all correspondence to: m.eldiasty@squ.edu.om

References

[1] Axelsson P. DEM generation from laser scanner data using adaptive TIN models. International Archives of Photogrammetry and Remote Sensing. 2000;**33**(B4/1; PART 4):111-118

[2] Vincent G, Sabatier D, Blanc L, Chave J, Weissenbacher E, Pélissier R, et al. Accuracy of small footprint airborne LiDAR in its predictions of tropical moist Forest stand structure. Remote Sensing of Environment. 2012;**125**:23-33. DOI: 10.1016/j.rse.2012.06.019

[3] Nie S, Wang C, Dong P, Xiaohuan X, Luo S, Qin H. A revised progressive TIN densification for filtering airborne LiDAR data. Measurement. 2017;**104**:70-77. DOI: 10.1016/j.measurement.2017.03.007

[4] Shi X, Hongchao M, Chen Y, Zhang L, Zhou W. A parameter-free progressive TIN densification filtering algorithm for Lidar point clouds. International Journal of Remote Sensing. 2018;**4**:1-14. DOI: 10.1080/01431161.2018.1468109

[5] Sithole G. Filtering of laser altimetry data using slope adaptive filter. International Archives of Photogrammetry & Remote Sensing. 2001;**34**(3/W4):203-210

[6] Vosselman G. Slope based filtering of laser altimetry data. International Archives of Photogrammetry and Remote Sensing. 2000;**33**(B3/2; PART 3):935-942

[7] Susaki J. Adaptive slope filtering of airborne LiDAR data in urban areas for digital terrain model (DTM) generation. Remote Sensing. 2012;**4**(6):1804. DOI: 10.3390/rs4061804

[8] Kilian J, Haala N, Englich M. Capture Andevaluation of Airborne Laser Scanner Data. International Archives of Photogrammetry & Remote Sensing; 1996

[9] Zhang K, Chen SC, Whitman D, Shyu ML. A progressive morphological filter for removing nonground measurements from airborne LIDAR data. IEEE Transactions on Geoscience & Remote Sensing. 2003;**41**(4):872-882. DOI: 10.1109/TGRS.2003.810682

[10] Mongus D, Borut Ž. Parameter-free ground filtering of LiDAR data for automatic DTM generation. ISPRS Journal of Photogrammetry and Remote Sensing. 2012;**67**:1-12. DOI: 10.1016/j.isprsjprs.2011.10.002

[11] Kobler A, Pfeifer N, Ogrinc P, Todorovski L, Oštir K, Sašo D. Repetitive interpolation: A robust algorithm for DTM generation from aerial laser scanner data in forested terrain. Remote Sensing of Environment. 2007;**108**(1):9-23. DOI: 10.1016/j.rse.2006.10.013

[12] Polat N, Uysal M, Toprak AS. An investigation of DEM generation process based on LiDAR data filtering, decimation, and interpolation methods for an urban area. Measurement. 2015;**75**:50-56. DOI: 10.1016/j.measurement.2015.08.008

[13] Shi W, Zheng S, Tian Y. Adaptive mapped least squares SVM-based smooth fitting method for DSM generation of LIDAR data. International Journal of Remote Sensing. 2009;**30**(21):5669-5683. DOI: 10.1080/01431160802709237

[14] Hu X, Yuan Y. Deep-learning-based classification for DTM extraction from

ALS point cloud. Remote Sensing. 2016;**8** (9):730. DOI: 10.3390/rs8090730

[15] Cai Z, Ma H, Zhang L. Feature selection for airborne LiDAR data filtering: A mutual information method with Parzon window optimization. GIScience & Remote Sensing. 2020;**57** (3):323-337. DOI: 10.1080/15481603.2019.1695406

[16] Jolliffe IT. Principal Component Analysis. 2nd ed. New York, NY: Springer; 2002

[17] Ghilani C. Adjustment Computations: Spatial Data Analysis. Sixth ed. John Wiley & Sons, Inc; 2017

Section 2

Machine Learning for Time Series Analysis

Chapter 2

ARIMA Models with Time-Dependent Coefficients: Official Statistics Examples

Guy Mélard

Abstract

About 25 years ago, effective methods for dealing with time series models that vary with time appeared in the statistical literature. Except in a few cases, they have never been used for economic statistics. In this chapter, we consider autoregressive integrated moving average (ARIMA) models with time-dependent coefficients (tdARIMA) applied to monthly industrial production series. We start with a small-size study with time-dependent integrated autoregressive (tdARI) models on Belgian series compared to standard ARI models with constant coefficients. Then, a second, bigger, illustration is given on 293 U.S. industrial production time series with tdARIMA models. We employ the software package Tramo to obtain linearized series and model specifications and build both the ARIMA models with constant coefficients (cARIMA) and the tdARIMA models, using specialized software. In these tdARIMA models, we use the simplest specification for each coefficient: a simple regression with respect to time. Surprisingly, for a large part of the series, there are statistically significant slopes, indicating that the tdARIMA models fit better the series than the cARIMA models.

Keywords: nonstationary process, time series, time-dependent model, time-varying model, local stationarity

1. Introduction

About 25 years ago, effective methods for dealing with time series models that vary with time appeared in the statistical literature. Except in a few cases, like Van Bellegem and von Sachs [1] for marginal heteroscedasticity in financial data or Kapetanios et al. [2], they are not used for economic statistics. In this chapter, we consider autoregressive integrated moving average (ARIMA) models with time-dependent coefficients (tdARIMA) that provide a natural alternative to standard ARMA models. Several theories appeared in the last 25 years for parametric estimation in that context, including Dahlhaus' approach based on locally stationary processes, see Dahlhaus [3, 4]. To simplify the presentation of the method in Section 2, we first focus on autoregressive integrated (ARI) models before going to the general case of ARIMA. Section 3 is devoted to illustrations of official time series, more precisely industrial production series. We start with a small-size study on Belgian monthly industrial production and show an improvement for time-dependent autoregressive integrated (tdARI) models with respect to standard ARI models with constant coefficients. Then, a second, bigger, illustration of tdARIMA models is given on 293 U.S. industrial production time series, already used by Proietti and Lütkepohl [5], with a different objective. We employ the software package Tramo from Gómez and Maravall [6] to obtain linearized series and model specifications, and we built both ARIMA models with constant coefficients (cARIMA) and tdARIMA models based on the Tramo specifications. This is done in specialized software since no existing package can cope with these tdARIMA models. In these tdARIMA models, we use the simplest specification for each coefficient: a simple regression with respect to time, hence two parameters, a constant and a slope. Indeed, this is the closest departure from constancy, and this seems natural in an evolving world. We will see that, for a large part of the series, there are statistically significant slopes, indicating that the tdARIMA models fit better the series than the cARIMA models. In the second step, since many of the slopes introduced as additional parameters in the model are not significantly different from 0, they are omitted one by one, starting with the least significant one, until all the remaining slopes are different from 0 at the 5% level. Most of the summary results are improved. Section 4 contains our conclusions.

2. Methods

We consider the well-known class of multiplicative seasonal ARIMA models, see e.g. Gómez and Maravall and Box *et al*. [6, 7]. Models with time-dependent coefficients appear often in econometrics but not in ARIMA models. For a very long time series, there is no reason that the coefficients would stay constant. They can be supposed to vary slowly with time although breaks could also be considered. This is the reason why linear (or other) functions of time replace the constant coefficients. Time series models with time-varying coefficients have been studied, mainly from a theoretical point of view. In addition to [3, 4], several papers [8–10] provide conditions for the asymptotic properties, hence the justification for statistical inference. Otherwise, our tests on slopes would have no foundation. These conditions are of course enforced in the estimation procedure.

2.1 The model

To illustrate a simple ARIMA model with a time-dependent coefficient, we can consider the ARMA(1,1) model. Let the series be denoted by $y = (y_1, y_2, \ldots, y_n)$. Then a tdARMA(1,1) model is described by the equation as follows:

$$y_t = \phi_t^{(n)} y_{t-1} + e_t - \theta_t^{(n)} e_{t-1}, \tag{1}$$

where the e_t are independent random variables with mean zero and with standard deviation σ, and the time-dependent coefficients $\phi_t^{(n)}$ and $\theta_t^{(n)}$ depend on time t, also on n, the length of the series, and also on a small number of parameters stored in a $m \times 1$ vector β. The simplest specification for $\phi_t^{(n)}$, for example, is as follows:

$$\phi_t^{(n)}(\beta) = \phi + \frac{1}{n-1}\left(t - \frac{n+1}{2}\right)\phi', \tag{2}$$

where ϕ is an intercept and ϕ' is a slope, and a similar expression for $\theta_t^{(n)}(\beta)$ using two other parameters θ and θ'. The vector β contains all parameters to be estimated, those in $\phi_t^{(n)}(\beta)$ (like ϕ and ϕ', here) and $\theta_t^{(n)}(\beta)$ (θ and θ'), but not the scale factor σ which is estimated separately. For the corresponding cARIMA model, there is of course no slope, i.e., $\phi' = \theta' = 0$. For a lag k instead of 1, we add a subscript k to the coefficient symbols.

Let us now consider a general tdARMA(p, q) model. It is defined by the equation

$$y_t = \sum_{k=1}^{p} \phi_{tk}^{(n)}(\beta) y_{t-k} + e_t - \sum_{k=1}^{q} \theta_{tk}^{(n)}(\beta) e_{t-k}, \tag{3}$$

where the coefficients $\phi_{tk}^{(n)}(\beta)$, $k = 1, \dots, p$, and $\theta_{tk}^{(n)}(\beta)$, $k = 1, \dots, q$, are deterministic functions of t and, possibly, of n. The e_t, $t = 1, 2, \dots$, are like before. We suppose that the additional number of parameters is small. Practically, for economic time series, linear or exponential functions of time, like in Eq. (2), seem to be enough instead of constant coefficients, but there is no problem to use other functions, up to some point. In other cases, see Alj *et al.* [11], periodic functions can be considered. In practice, we suppose that the coefficients are constant before the first observation.

Adding marginal heteroscedasticity should also be tried. Van Bellegem and von Sachs [1] had already shown the usefulness of a time-dependent variance. Indeed, there is no reason why the innovation standard deviation is constant. We replace e_{t-k}, $k = 0, 1, \dots, q$, in Eq. (3) with $g_{t-k}^{(n)}(\beta) e_{t-k}$, where $g_t^{(n)}(\beta)$ is a (strictly positive) deterministic function of t and, possibly, of n, depending on the parameters, so that the standard deviation becomes $g_t^{(n)}(\beta)\sigma > 0$. Adding $g_t^{(n)}(\beta)$ is also covered by Azrak and Mélard [8, 12]. In practice, we used an exponential function of time for $g_t^{(n)}(\beta)$.

Since the series are nonstationary, we need to consider also regular ∇ and seasonal differences ∇_s, where s is the seasonal period ($s = 12$, for monthly data), on the possibly square roots or log-transformed observations. Furthermore, the series is not seasonally adjusted, so the so-called seasonal multiplicative models of Box *et al.* [7] are also needed.

2.2 The estimation method

For any tdARIMA model, we can estimate the parameters by maximizing the logarithm of the Gaussian likelihood. Time Series Expert [13], and more precisely its computational engine ANSECH is used for that purpose. It is based on an exact algorithm for the computation of the Gaussian likelihood [14] and an implementation of a Levenberg–Marquardt nonlinear least-squares algorithm. Under some very general conditions [8, 12], it is shown that the quasi-maximum likelihood estimator $\hat{\beta}$ converges to the true value of β, and $\hat{\beta}$ is asymptotically normal, more precisely $\sqrt{n}(\hat{\beta} - \beta) \xrightarrow{D} N(0, V^{-1})$, when $n \to \infty$ where $\xrightarrow{D}$ indicates convergence in distribution, and V^{-1} is the asymptotic covariance matrix. Moreover, V can be estimated as a by-product of estimation. Let us denote its estimator by $\hat{V}_n$. The Student t statistics shown in the next section make use of the standard errors deduced from the estimation of V. Using the asymptotic covariance matrix, it is also possible to design a Wald test for a subset **b** of r among the m parameters in β, for example, to test that all the slopes are equal to 0, using a χ^2 distribution. Let **R** be a $r \times m$ restriction matrix composed of the rows of the $m \times m$ identity matrix that correspond to the parameters in the subset **b**. Then, $b = R\beta$. The Wald statistic for testing that $b = 0$ is then $n\hat{b}'\left(R\hat{V}_nR'\right)^{-1}\hat{b}$, where $\hat{b}$ is the estimate of b and $'$ indicates transposition. Under the null hypothesis, the statistic converges in distribution to a χ^2 distribution with r degrees of freedom when $n \to \infty$.

Note that centering of time around its mean $(n + 1)/2$ in Eq. (2) improves the statistical properties of the estimators by reducing the amount of correlation between their elements and that the factor $1/(n–1)$ is there to avoid explosive behavior when $n \to \infty$.

Note also that the conditions for convergence and asymptotic normality are satisfied in the present case because a sufficient condition [15] is that the AR and MA polynomials have their roots outside the unit circle at all times and that condition is checked during estimation.

An asymptotic theory for locally stationary processes due to Dahlhaus [3, 4] can also be used. There seems to exist only one software implementation, the R package LSTS (for Locally Stationary Time Series) by Olea *et al.* [16] to support the estimation of locally stationary ARMA models, see also Palma *et al.* [17]. Since it does not cope with the multiplicative seasonal models necessary to deal with seasonally unadjusted time series, we have preferred to use Azrak and Mélard [8] with Time Series Expert for estimation. See Azrak and Mélard [18] for a comparison of the existing theories.

2.3 The datasets

In the first empirical analysis, the number of series is limited, and simple pure autoregressive models are used. The purpose is to show the basic elements of the methodology. We used a dataset of indices for the monthly Belgian industrial production for the period 1985–1994 by the various branches of activity, 26 in all. Nine years are used for fitting the models and a tenth year is used to compute ex-post forecasts and the mean absolute percentage error (MAPE). An automatic procedure is applied to fit ARIMA models and we retained the 20 series out of 26 for which pure integrated autoregressive or ARI$(p, d)(P, D)_{12}$ models are fitted to the series of 108 observations. Let us remind that these models are defined by the equation as follows:

$$\phi_p(L)\Phi_P(L^s)\nabla^d\nabla_{12}^D y_t = e_t, \quad (4)$$

where L is the lag operator, such that $Ly_t = y_{t-1}$, $\phi_p(L)$ and $\Phi_P(L^s)$ are, respectively, the regular autoregressive and the seasonal autoregressive polynomials, of degree p and $12P$ in L. The model can include transformations and interventions (additive or on the differenced series) which are not detailed here. The fit is characterized by the value of the SBIC criterion. For using time-dependent ARI, or tdARI, models, slope parameters are added for each of the existing coefficients, like ϕ' for ϕ in Eq. (2). The models have therefore coefficients that are linear functions of time. For models in multiplicative seasonal form, the product of the regular and seasonal polynomials is first computed and slope parameters are added to each lag, but only to lags smaller than 14, for practical reasons. For example, for the $AR(2)(1)_{12}$ model, with the polynomial in the lag operator L

$$\left(1-\phi_1 L-\phi_2 L^2\right)\left(1-\Phi_1 L^{12}\right)=\left(1-\phi_1 L-\phi_2 L^2-\Phi_1 L^{12}+\phi_1\Phi_1 L^{13}+\phi_2\Phi_1 L^{14}\right), \tag{5}$$

the specification is $\left(1-\phi_{t1}^{(n)}L-\phi_{t2}^{(n)}L^2-\phi_{t,12}^{(n)}L^{12}-\phi_{t,13}^{(n)}L^{13}+\phi_2\Phi_1 L^{14}\right)$, where $\phi_{t1}^{(n)}$ is like in Eq. (2), and

$$\phi_{t2}^{(n)}=\phi_2+\frac{1}{n-1}\left(t-\frac{n+1}{2}\right)\phi_2', \phi_{t,12}^{(n)}=\Phi_1+\frac{1}{n-1}\left(t-\frac{n+1}{2}\right)\phi_{12}',$$
$$\phi_{t,13}^{(n)}=-\phi_1\Phi_1+\frac{1}{n-1}\left(t-\frac{n+1}{2}\right)\phi_{13}',$$

with seven parameters instead of the full form $\left(1-\phi_{t1}^{(n)}B-\phi_{t2}^{(n)}B^2-\phi_{t,12}^{(n)}B^{12}-\phi_{t,13}^{(n)}B^{13}-\phi_{t,14}^{(n)}B^{14}\right)$ that would involve 10 parameters in all. This is enforced to restrict the number of parameters and avoid numerical problems. Note that the factor $1/(n-1)$ is there only for the asymptotic theory and will be omitted in practice.

In the second empirical analysis, we use a big dataset of U.S. industrial production time series, already used by Proietti and Lütkepohl [5] for assessing transformations in forecasting. See http://www.federalreserve.gov/releases/g17/ipdisk/ip_nsa.txt. These are now 293 time series from January 1986 to December 2018 at least. Most series start before and they are even a few ones starting in 1919. The models were fitted until December 2016 leaving the remaining months to compare the data to the ex-post forecasts, using either a fixed forecast origin for several horizons or rolling forecasts each for given horizons.

We employ the software package Tramo described by Gómez and Maravall [6] to obtain partially linearized series by removing outliers and trading day effects. Indeed, the presence of outliers and trading day effects can distort the analysis, as could be seen in preliminary analyses. Selecting the cARIMA models in an automated way is also done using Tramo. Then we replace the constant coefficients by linear functions of t for order $k \leq 13$, giving tdARIMA models, like in Eq. (2) for each lag k coefficient in the model. At this stage, we do not omit nonsignificant parameters. The cARIMA and tdARIMA models are fitted using the same specialized software package ANSECH included in Time Series Expert, to facilitate the comparison. See **Figure 1** for a schematic representation of the whole automatic procedure. For more complex time dependency, an automatic selection procedure like the one exposed by Van Bellegem and Dahlhaus [19] is possible.

We compare the results of tdARIMA versus cARIMA models using the following criteria:

- Is the highest t statistic of the td parameters, the slopes, in absolute value, larger than 1.96?
- Is the p-value of the global χ^2 statistic for the Wald test on the slopes smaller than 0.05?

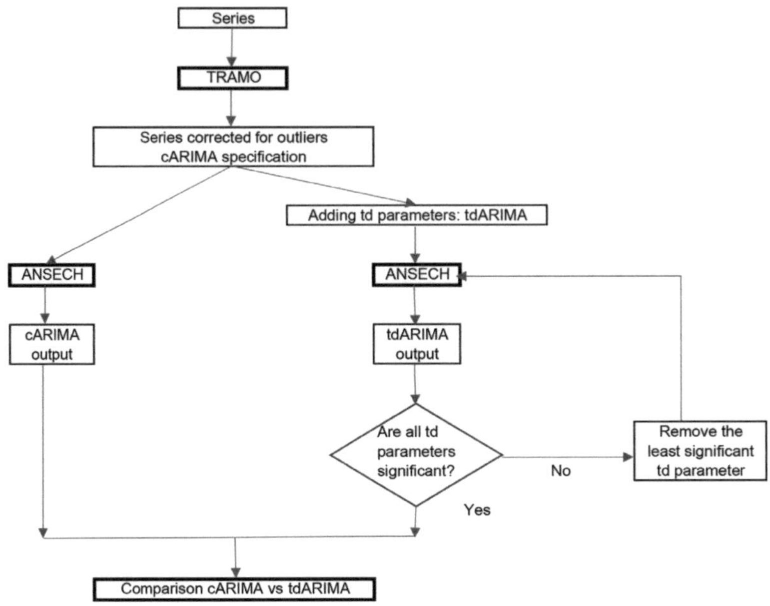

Figure 1.
Schematic representation for the whole automatic treatment.

- Is tdARIMA SBIC smaller than the corresponding cARIMA SBIC?
- Is tdARIMA residual standard deviation smaller than the corresponding cARIMA one?
- Is the tdARIMA *P*-value of the Ljung-Box (LB) statistic for residual autocorrelation (with lag 48) larger than the corresponding cARIMA one?
- Is tdARIMA MAPE in percent for 1 year (2017) and all horizons from 1 to 12 smaller than the corresponding cARIMA one?
- similarly, for rolling forecasts from December 2016, for several horizons 1, 3, 6, and 12, are tdARIMA MAPE in percent smaller than the corresponding cARIMA one?

In the early stage of this project, the data were limited to 2016 and without a correction for outliers or trading days, and only fixed origin forecasts were considered. This gave worse results that were indicative, and not conclusive.

Note that one can object against the use of the Ljung-Box test statistic to compare models, especially here because there is no foundation to its limit behavior for tdARIMA models. Like the other criteria, we use it as a descriptive indicator.

3. Empirical results

3.1 Two examples

Before showing the results, we will consider two examples, to justify the recourse to the class of tdARIMA models which is the object of this chapter.

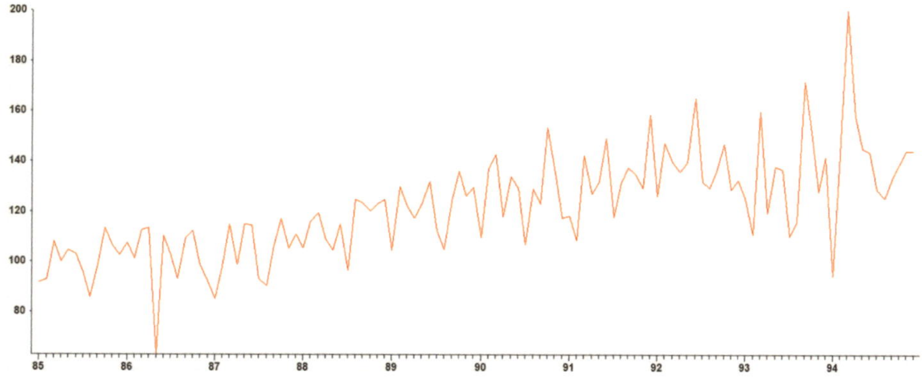

Figure 2.
The original series, the index of land transportation (TRTER).

The first example is taken from the first dataset: the index of land transportation (series TRTER) for Belgium for the period from January 1985 to December 1994, see **Figure 2**. Two additive interventions were automatically considered, respectively, in May 1986 and in February 1992. Let I8605 and I9202 denote, respectively, the corresponding binary variables. Otherwise, the series is taken in square roots and seasonally differenced. Let us denote the transformed series TRTER_TF after all these operations. It equal to $\nabla_{12}\left(\sqrt{\text{TRTER}} - b_{\text{I8605}}\text{I8605} - b_{\text{I9202}}\text{I9202}\right)$, up to a normalizing factor, and shown in **Figure 3**. The partial autocorrelations of that series shown in **Figure 4** show a truncation after lag 12. The usual Box and Jenkins analysis lead to the suggestion of a seasonal AR model. Then adding time dependency to the AR coefficient leads to the following model:

$$(1 - \Phi_{t,12}\text{B})\left[\nabla_{12}\left(\sqrt{\text{TRTER}} - b_{\text{I8605}}\text{I8605} - b_{\text{I9202}}\text{I9202}\right) - \mu\right] = e_t,$$

where the $\Phi_{t,12}$, is estimated by $-0.686 + 6.47\,10^{-03}\,(t - 60.5)$, and the estimates of b_{I8605}, b_{I9202}, and μ are, respectively, equal to -30.9, 24.8, and 4.23. The standard error corresponding to the slope of $\Phi_{t,12}$ is equal to $1.70\,10^{-03}$, so the associated Student statistic is equal to 3.8, hence the slope is significantly different from 0. To explain that significance, let us look at the partial autocorrelation at lag 12 for the transformed series TRTER_TF: for the first 4 years it is -0.488 and for the last 4 years it is -0.391. That explains the significantly positive slope for $\Phi_{t,12}$

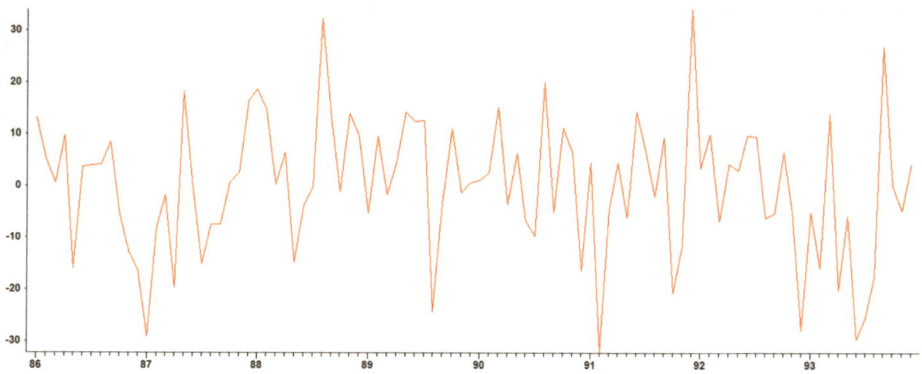

Figure 3.
The transformed series (TRTER_TF).

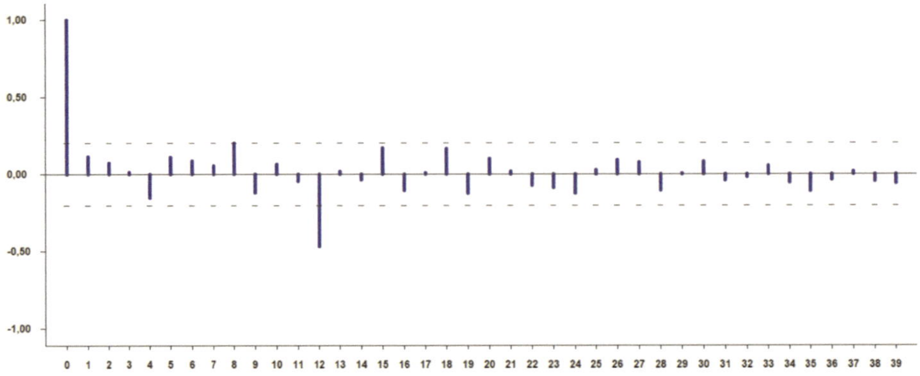

Figure 4.
The partial autocorrelation function of the transformed series.

The second example is taken from the second dataset: the U.S. production index of clothing (B51212) for the period from January 1986 to January 2019, see **Figure 5**.

Tramo has adjusted the series for outliers and proposed a logarithmic transform, and both a regular and a seasonal difference, giving the transformed series BS51212DS shown in **Figure 6**.

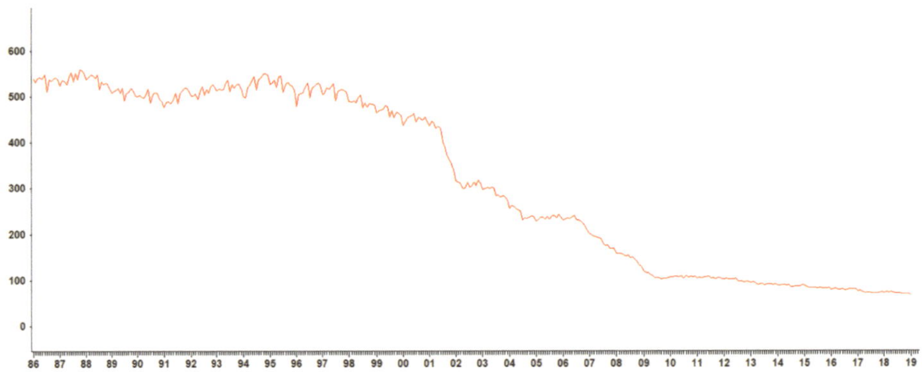

Figure 5.
The original series, the U.S. production index of clothing (B51212).

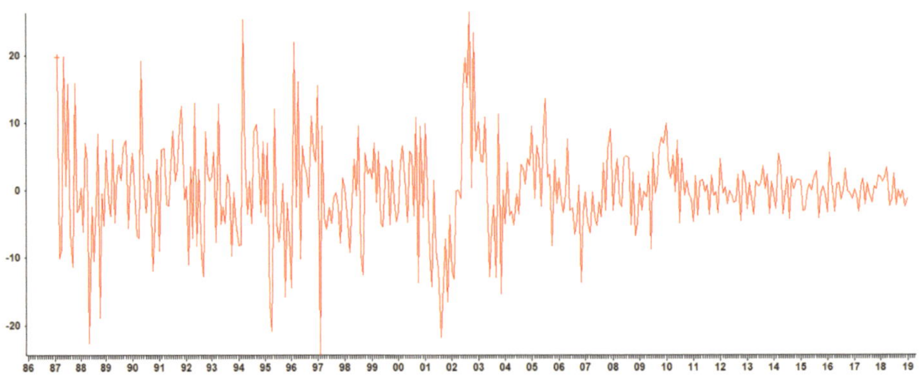

Figure 6.
The transformed B51212 series (B51212DS).

Subsequently, Tramo has suggested modeling the series by a seasonal ARIMA model with a regular autoregressive polynomial of degree 3 and a seasonal moving average. We fitted that model using ANSECH and obtained

$$(1 + 0.035L - 0.142L^2 - 0.249L^3)\nabla\nabla_{12} \log (\text{B51212}_t) = (1 - 0.850L^{12})e_t.$$

Then, we replaced the constant coefficients with linear functions of time and replaced the constant innovation variance with an exponential function of time. Omitting one by one the nonsignificant parameters at the 5% probability level, we obtained finally the following heteroscedastic model but with constant autoregressive and moving average polynomials:

$$(1 + 0.041L - 0.142L^2 - 0.228L^3)\nabla\nabla_{12} \log (\text{B51212}_t) = (1 - 0.855L^{12})g_t e_t,$$

where g_t is given by $g_t = \exp(0.001753(t - 193))$.

3.2 First empirical analysis

In the first experiment, the number of series is small and simple pure integrated autoregressive models are used.

Table 1 shows the main results, including those tdAR coefficients for which the test of zero slope leads to a rejection at the 5% level, and the corresponding t-statistic.

For example, the $\text{AR}(2)(1)_{12}$ model in Eq. (5) which was shown in Section 2.3 is used for nonmetallic manufacturing, in addition to a regular and a seasonal difference. In that case, the standard ARI model is better than the tdARI model for SBIC (826 versus 830) and provides also betters forecasts (MAPE = 5.5% versus 9.4%) although there is one significant slope for lag 12 with a Student statistic of 3.6. Even if tdARI models are not systematically better, they often produce better forecasts and sometimes show a better fit or at least some statistically significant slope parameters at the 5% level. All the nonsignificant slopes were left in the model and that can explain why the SBIC criterion was generally worse for tdARI models. Since that analysis was promising, we were led to consider a bigger dataset.

3.3 Second empirical analysis

The second empirical analysis bears on 293 seasonally unadjusted time series in a dataset of U.S. industrial production. We will show three tables of results. **Table 2** presents a summary of the dataset resulting from Tramo. For example for 280 series out of 293, a regular difference was used, which accounts for 96% of the dataset. As said above, we preserved this and all parameters in our cARIMA and tdARIMA models. For the tdARIMA models, slopes were added to all autoregressive and moving average coefficients for lag less or equal to 13.

Table 3 is based on the initial tdARIMA models, with possibly nonsignificant parameters. We show the percentages for each criterion across the 293 series. For example, more than 50% of the dataset had at least one of the slopes with a Student t value greater than 1.96. If we use the Wald test, which should show a better view, for more than 44% of the series, the hypothesis of null slopes leads to a rejection at the 5% level. If the series were randomly drawn from cARIMA processes, we should expect 5% of rejections, on average. Of course, because of the multiple-test argument, the

Branch (Name)	Orders	ARI	tdARI		
	(*p*, *d*) **(*P*, *D*)**	**SBIC** ***MAPE***	**SBIC** ***MAPE***	**parameter**	***t*-Value**
Food, beverages (ALIBOR)	(3, 0) (0, 1)	**655** *4.0*	669 ***3.9***	none	
Other extraction of minerals (AUEXTR)	(2, 0) (0, 1)	**904** *10.5*	910 ***9.4***	AR2	2.8
Wood-processing, furniture (BOIME)	(3, 1) (1, 1)	**778** *6.4*	792 ***5.9***	AR12	2.6
Hosiery (BONNE)	(2, 0) (1, 1)	702 *9.1*	**692** ***5.5***	AR1 AR2 AR12	2.9 8.7 11.0
Commerce (COMME)	(3, 0) (1, 1)	**553** ***1.7***	557 *2.4*	AR12	3.7
Construction (CONST)	(3, 0) (0, 1)	**827** ***8.0***	830 *23.6*	AR2	2.8
Petrol derivatives (DERPE)	(1, 1) (0, 0)	**888** *5.2*	891 *5.2*	none	
Petrol distribution (DISPE)	(1, 0) (1, 0)	**941** *10.8*	951 ***10.7***	none	
Metal processing (FABME)	(2, 0) (0, 1)	682 ***6.6***	**681** *7.2*	AR2	−2.1
Manufacture of textiles (FILAT)	(2, 1) (0, 1)	**739** ***5.5***	749 *6.4*	none	
Gas production/distribution (GAZ)	(3, 0) (1, 1)	897 *2.1*	**891** ***2.0***	AR2 AR3	−2.8 −4.5
Construction materials (MATCO)	(1, 1) (1, 1)	809 ***4.3***	**807** *5.7*	AR13	−3.3
Nonmetallic manufacturing (NONFE)	(2, 1) (1, 1)	**826** ***5.5***	830 *9.4*	AR12	3.6
Paper/paperboard industry (PAPCA)	(3, 0) (1, 1)	**746** *5.9*	758 ***5.8***	AR2	−2.1
Iron and steel (SIDER)	(3, 0) (1, 1)	**833** ***5.9***	836 *9.2*	AR1 AR12	4.0 3.6
Manufacture of tobacco (TABAC)	(3, 0) (1, 1)	**778** ***10.9***	791 *13.2*	AR3 AR12	3.1 5.4
Aviation (TRAER)	(3, 1) (1, 1)	**746** *8.8*	753 ***8.7***	AR12	6.3
Maritime transportation (TRMAR)	(2, 1) (1, 1)	**732** *3.6*	740 ***2.6***	AR12 AR13	2.3 3.6
Land transportation (TRTER)	(0, 0) (1, 1)	854 ***12.1***	**849** *12.5*	AR12	−2.6
Manufacture of clothing (VETEM)	(2, 1) (0, 1)	**757** *26.7*	767 *26.7*	none	

Table 1.
For each branch of the economy, we give the orders (p,d)(P,D) of the model, SBIC *and* MAPE *(in italics) for the raw ARI model and for the tdARI model (results in bold type are better), the statistically significant slopes (ARk denotes* ϕ'_k*) and the corresponding* t*-value.*

Specification	#series	%	Note
Levels vs. logs	70/223	24/76	
Regular diff. vs. none	280/13	96/4	4% with 2 differences
Seasonal diff. vs. none	279/14	95/5	0% with 2 differences
Stationary vs. nonstationary	0/293	0/100	
Airline model vs. other	84/209	29/71	
Outliers vs. none		79/21	0–20, on average 2.63
Trading day effect vs. none		47/53	
Easter effect vs. none		40/60	
ARMA parameters vs. none		100/0	1–7, on average 3.12

Table 2.
Summary of the model selections made by Tramo on the 293 U.S. industrial production series.

Student tests on the slopes would give a higher proportion of rejections. A simulation study will confirm this later. The use of the Wald test in the present context is therefore essential. Some results are partially deceptive but can be explained: only about 4% of the series have a smaller SBIC for the tdARIMA but this is mainly due to the useless parameters. About one-half of the series have a smaller residual standard deviation, but for more than 57% the test on residual autocorrelation, based on the Ljung-Box test with 48 lags, gives a better result.

If we retain only the series where the Wald test rejected the constancy of the coefficients, the percentage of smaller SBIC for tdARIMA models is only slightly higher at about 9% and reaches 61% for the residual standard deviation. The percentage for the Ljung-Box test is lower. Indeed, the theory for that test was never undertaken for tdARIMA models. Forecasting performance was evaluated using the MAPE criterion. For fixed origin forecasts, about 47% of the series have a smaller MAPE for the tdARIMA models rather than for the cARIMA models. Among the series for which

Criteria	Percentage	Notes	Percentage if td significant
Highest $\|t\|$ statistic of td parameters >1.96	50.17	(*)	
p-value of global test of stationarity <0.05	44.71		100.00
tdARIMA SBIC < cARIMA SBIC	04.10	(**)	9.16
tdARIMA residual std. dev < cARIMA	49.15		61.83
tdARIMA LB P-value > cARIMA	57.68		54.20
tdARIMA forecasting MAPE < cARIMA	47.44		45.04
tdARIMA h = 1 rolling forecasts MAPE < cARIMA	32.76		31.30
tdARIMA h = 3 rolling forecasts MAPE < cARIMA	32.42		29.01
tdARIMA h = 6 rolling forecasts MAPE < cARIMA	37.88		40.46
tdARIMA h = 12 rolling forecasts MAPE < cARIMA	36.86		37.40

Notes: (*) Statistically significant slope parameters at the 5% level, (**) Nonsignificant parameters were not omitted.

Table 3.
For each criterion, the percentages of improvement from cARIMA models to tdARIMA models are given over the 293 U.S. series.

time dependency is retained, only 45% of them benefit from better forecasts. For rolling forecasts for various horizons, the percentages are even smaller, in particular for horizons of 1 and 3 months. The percentages are about the same if the Wald test rejects constancy or not. That means that, even if the introduction of time-dependency improved the fits, it does not improve the forecasts. Let us remind that, at this stage, the tdARIMA may have many statistically nonsignificant slopes.

For **Table 4**, starting from the full tdARIMA models of **Table 3**, we omitted, one by one, the most nonsignificant slope at the 5% level, see **Figure 1**. In the end, all remaining slopes are thus significantly different from 0. This was done in an automated way in order to avoid mistakes. We will refer to these models as parsimonious tdARIMA models. Of course, the cARIMA models are the same as previously, essentially the same as given by Tramo, but estimated with more digits of accuracy. We notice that the percentage of at least one statistically significant slope, 54.61%, differs slightly from the percentage of rejection of the Wald test on all the slopes, 54.27%. Indeed, for one series (G325A4, Chemicals except for pharmaceuticals and medicines), there are two slightly significant slopes but the global test does not reject their nullity, although the p-value is close to 0.05. Anyway, these percentages of improved tdARIMA models are slightly higher than in **Table 3**.

The fitting results are partially better with more than 18% smaller SBIC for tdARIMA models (respectively 34% if we condition on the rejection of the Wald test). Some are worse, however, with 38% for the residual standard deviation instead of 49% for the fully parameterized model (respectively 70% and 61%, if we condition on the rejection of the Wald test), and 27% for the residual autocorrelation instead of 57% for the full model (respectively 50% and 54%, if we condition on the rejection of the Wald test).

Strangely, the forecasting performance with a fixed origin is worse for the parsimonious model than for the full model with the percentage of improvement of tdARIMA models with respect to cARIMA models equal to 28%, instead of 47% (respectively 51% and 45%, if we condition on the rejection of the Wald test). That means that the omitted slopes seem to contribute to the forecasting performance but

Criteria	Percentage	Notes	Percentage if td significant
Highest $\|t\|$ statistic of td parameters >1.96	54.61	(*)	
p-value of global test of stationarity <0.05	54.27		100.00
tdARIMA SBIC < cARIMA SBIC	18.77	(**)	34.59
tdARIMA residual std. dev < cARIMA	38.23		70.44
tdARIMA LB P-value > cARIMA	27.65		50.31
tdARIMA forecasting MAPE < cARIMA	28.33		51.57
tdARIMA h = 1 rolling forecasts MAPE < cARIMA	20.48		37.74
tdARIMA h = 3 rolling forecasts MAPE < cARIMA	17.75		32.70
tdARIMA h = 6 rolling forecasts MAPE < cARIMA	18.77		34.59
tdARIMA h = 12 rolling forecasts MAPE < cARIMA	21.84		40.25

Notes: (*) Statistically significant slope parameters at the 5% level, (**) Contrarily to ***Table 3****, nonsignificant slope parameters are omitted one by one until all were statistically significant.*

Table 4.
For each criterion, the percentages of improvement from cARIMA models to tdARIMA models are given over the 293 U.S. series. The last column contains percentages conditional to the rejection of nullity of all the slopes by the Wald test.

that, among the series with time-dependent coefficients, about one-half have provided better forecasts. The picture for rolling forecasts is again worse for the parsimonious models with smaller percentages of improvement in the range of 17-22%, according to the horizon, instead of 32-37% for the full models, but again similar under the condition of rejection of the Wald test (respectively 32-44% instead of 29-40%). Surprisingly, the percentages are systematically higher for horizons 6 and 12 months rather than for those of 1 and 3 months.

One can object that introducing time dependency can introduce some over-fitting: a certain proportion of the tests of nullity of the slopes ϕ'_k or θ'_k can lead to rejection, about 5% when there is only one slope, more otherwise.

To try to answer that natural question, we generated artificially 320 series using cARIMA models, with the same length of 372, again leaving the last 12 values. We have used an airline model for that purpose instead of the large variety of models fitted by Tramo-Seats. Then we added time dependency and proceeded exactly like before. The results are shown in **Table 5**. The percentage of 14.06 for the first criterion (instead of five) shows that our rough examination of the largest $|t|$ value should be better replaced by a simultaneous test on the td parameters, as we did. For SBIC, there are many superfluous parameters, as could be guessed. But about one-half of the tdARIMA models give smaller residual standard deviations, less residual autocorrelation, and smaller forecast errors than their cARIMA counterparts, as expected.

The results show that for a majority of series there is (i) at least one statistically significant slope parameter at the 5% level, (ii) rejection of the nullity of all the slopes using a Wald test that provides better-founded results than the t-tests, (iii) smaller residual standard deviation, and (iv) less residual autocorrelation. This is true for the full tdARIMA model specifications but also, at least partly, with more parsimonious tdARIMA models obtained by omitting, one by one, the statistically nonsignificant slopes. At least it is true conditionally on significant time dependency, i.e. when the Wald test rejects the constancy of the coefficients.

The results for the SBIC criterion are not good. For the full tdARIMA model, an explanation is the presence of nonsignificant slope parameters. It remains, however, for the parsimonious models. The only unsatisfactory aspect of tdARIMA models is that they fail to improve the forecasts for a majority of the series. Indeed, they confirm that only one-third of the "time-dependent series", i.e. those series which have at least one statistically significant slope parameter, provide better forecasts with a tdARIMA model than with a cARIMA model.

We had already observed similar results with slightly shorter series of the big dataset when the outliers and trading day effects were not handled. Consequently, the

Criteria	Percentage Artificial series	Percentage U.S series
Highest $|t|$ statistic of td parameters >1.96	14.06	50.17
tdARIMA SBIC < cARIMA SBIC	00.00	04.10
tdARIMA residual std. dev < cARIMA	42.81	49.15
tdARIMA LB P-value > cARIMA	43.75	57.68
tdARIMA forecasting MAPE < cARIMA	47.81	47.44

Table 5.
For each criterion, the percentages of improvement from going from cARIMA models to tdARIMA models are given over the 320 artificial series. The last column contains the corresponding percentages obtained for the U.S. series taken from ***Table 3****.*

presence of outliers or trading day effects is not the cause of better fits by tdARIMA models, as we feared. A common feature is nevertheless that the forecasts are not better by replacing the cARIMA models with tdARIMA models. This is surprising although we know that a better fit is not a guarantee for better forecasts. It should be investigated why the forecasts seem to be worse in the U.S. series for tdARIMA models. Of course, it can be due to a global change in 2016.

4. Conclusions

It took several decades to go from ARIMA models with constant coefficients to suitable and powerful generalizations with time-dependent coefficients that vary deterministically. We showed the usefulness of the approach for dealing with official statistics time series that have generally a seasonal component.

We used linear functions of time. We do not hope that other functions than linear functions should be useful with the inconvenience to add many parameters, except if we exploit the fact that since 2019 most of the series in the dataset are available before 1986, often since 1972, or sometimes earlier.

Finally, one weak point in the analysis is due to the detection of outliers and trading day effects, and the time series linearization is based on cARIMA models. If the time dependency of the coefficients becomes serious for very long official time series, it would be worth trying to extend Tramo features to tdARIMA models, e.g. to detect outliers simultaneously with the estimation of time-dependent coefficients for the ARIMA model.

On the other side, it would be also interesting to conclude that traditional cARIMA models are enough to forecast very long time series and that no substantial gain can be obtained by considering tdARIMA models.

It can be interesting to repeat the analysis with other datasets, quarterly or preferably monthly, like those maintained by Eurostat. Good candidates would be in the industry, trade, and services, short-term business statistics, production, turnover, etc. A U.S. database like FRED (https://research.stlouisfed.org/econ/mccracken/fred-databases/) could also be considered.

Acknowledgements

I thank Agustín Maravall (for his help in producing linearized time series in Tramo), Rajae Azrak (for her contributions to the theory), Ahmed Ben Amara (for his contribution to a very first version of a part of the program chain which includes Tramo-Seats and Microsoft Excel Visual Basic modules, in addition to our specialized code for estimating tdARIMA models), and Dario Buono, Team Leader of Methodology, Eurostat, Unit B1 (for exchanges, suggestions, and encouragement on previous versions of this chapter). Finally, I thank the editors for their remarks and Mrs. Karla Skuliber, the author service manager, for her efficiency.

Conflict of interest

I declare there is no conflict of interest.

Author details

Guy Mélard
Université libre de Bruxelles, ECARES, Brussels, Belgium

*Address all correspondence to: guy.melard@ulb.be

References

[1] Van Bellegem S, von Sachs R. Forecasting economic time series with unconditional time-varying variance. International Journal of Forecasting. 2004;**20**:611-627. DOI: 10.1016/j.ijforecast.2003.10.002

[2] Kapetanios G, Marcellino M, Venditti F. Large time-varying parameter VARs: A nonparametric approach. Journal of Applied Econometrics. 2019;**34**:1027-1049. DOI: 10.1002/jae.2722

[3] Dahlhaus R. Maximum likelihood estimation and model selection for locally stationary processes. Journal of Nonparametric Statistics. 1996;**6**: 171-191. DOI: 10.1080/10485259608832670

[4] Dahlhaus R. Fitting time series models to nonstationary processes. Annals of Statistics. 1997;**25**:1-37. DOI: 10.1214/aos/1034276620

[5] Proietti T, Lütkepohl H. Does the Box-Cox transformation help in forecasting macroeconomic time series? International Journal of Forecasting. 2013;**29**:88-99. DOI: 10.1016/j.ijforecast.2012.06.001

[6] Gómez V, Maravall A. Automatic modelling methods for univariate series. In: Peña D, Tiao GC, Tsay RS, editors. A Course in Time Series Analysis. New York: Wiley; 2001. pp. 171-201. DOI: 10.1002/9781118032978.ch7

[7] Box GEP, Jenkins GM, Reinsel GC, Ljung GS. Time Series Analysis, Forecasting and Control. 5th ed. New York: Wiley; 2015 xxvi+669 p

[8] Azrak R, Mélard G. Asymptotic properties of quasi-maximum likelihood estimators for ARMA models with time-dependent coefficients. Statistical Inference for Stochastic Processes. 2006; **9**:279-330. DOI: 10.1007/s11203-005-1055-6

[9] Dahlhaus R. A likelihood approximation for locally stationary processes. Annals of Statistics. 2000;**28**: 1762-1794. DOI: 10.1214/aos/1015957480

[10] Dahlhaus R. Locally stationary processes. In: Subba Rao T, Subba Rao S, Rao CR, editors. Handbook of Statistics, Volume 30: Time Series Analysis: Methods and Applications. Amsterdam: Elsevier; 2012. pp. 145-159. DOI: 10.1016/B978-0-444-53858-1.00013-2

[11] Alj A, Azrak R, Ley C, Mélard G. Asymptotic properties of QML estimators for VARMA models with time-dependent coefficients. Scandinavian Journal of Statistics. 2017; **44**:617-635. DOI: 10.1111/sjos.12268

[12] Azrak R, Mélard G. Asymptotic properties of conditional least-squares estimators for array time series. Statistical Inference for Stochastic Processes. 2021;**24**:525-547. DOI: 10.1007/s11203-021-09242-8

[13] Mélard G, Pasteels J-M. User's Manual of Time Series Expert (TSE Version 2.3). Brussels: Institut de Statistique, Université Libre de Bruxelles; 1998. Available from: https://dipot.ulb.ac.be/dspace/retrieve/829842/TSE23E.PDF [Accessed: October 17, 2022]

[14] Mélard G. The likelihood function of a time-dependent ARMA model. In: Anderson OD, Perryman MR, editors. Applied Time Series Analysis. Amsterdam: North-Holland; 1982. pp. 229-239

[15] Mélard G. An indirect proof for the asymptotic properties of VARMA model estimators. Econometrics and Statistics. 2022;**21**:96-111. DOI: 10.1016/j.ecosta.2020.12.004

[16] Olea R, Palma W, Rubio P. Package LSTS. 2015. Available from: https://cran.r-project.org/web/packages/LSTS/LSTS.pdf [Accessed: October 17, 2022]

[17] Palma W, Olea R, Ferreira G. Estimation and forecasting of locally stationary processes. Journal of Forecasting. 2013;**32**:86-96. DOI: 10.1002/for.1259

[18] Azrak R, Mélard G. Autoregressive models with time-dependent coefficients - a comparison between several approaches. Stats. 2022;**5**:784-804. DOI: 10.3390/stats5030046

[19] Van Bellegem S, Dahlhaus R. Semiparametric estimation by model selection for locally stationary processes. Journal of the Royal Statistical Society Series B. 2006;**68**:721-746. DOI: 10.1111/j.1467-9868.2006.00564.x

Section 3

Time Series Methods

Chapter 3

Methods of Conditionally Optimal Forecasting for Stochastic Synergetic CALS Technologies

Igor N. Sinitsyn and Anatoly S. Shalamov

Abstract

Problems of optimal, sub- and conditionally optimal filtering and forecasting in product and staff subsystems at the background noise in synergistical organization-technical-economical systems (SOTES) are considered. Nowadays for highly available systems the problems of creation of basic systems engineering principles, approaches and information technologies (IT) for SOTES from modern spontaneous markets at the background inertially going world economics crisis, weakening global market relations at conditions of competition and counteraction reinforcement is very important. Big enterprises need IT due to essential local and systematic economic loss. It is necessary to form general approaches for stochastic processes and parameters estimation in SOTES at the background noises. The following notations are introduced: special observation SOTES (SOTES-O) with own organization-product resources and internal noise as information from special SOTES being enact noise (SOTES-N). Conception for SOTES structure for systems of technical, staff and financial support is developed. Linear, linear with parametric noises and nonlinear stochastic (discrete and hybrid) equations describing organization-production block (OPB) for three types of SOTES with their planning-economical estimating divisions are worked out. SOTES-O is described by two interconnected subsystems: state SOTES sensor and OPB supporting sensor with necessary resources. After short survey of modern modeling, sub- and conditionally optimal filtering and forecasting basic algorithms and IT for typical SOTES are given. Influence of OTES-N noise on rules and functional indexes of subsystems accompanying life cycle production, its filtration and forecasting is considered. Experimental software tools for modeling and forecasting of cost and technical readiness for parks of aircraft are developed.

Keywords: sub- and conditionally optimal filtering and forecasting (COF and COFc), continuous acquisition logic support (CALS), organizational-technical-economical systems (OTES), probability modeling, synergetical OTES (SOTES)

1. Introduction

Stochastic continuous acquisition logic support (CALS) is the basis of integrated logistic support (ILS) in the presence of noises and stochastic factors in organizational-technical-economic systems (OTES). Stochastic CALS methodology was firstly developed in [1–5]. According to contemporary notions in broad sense ILS being CALS basis represents the systems of scientific, design-project, organization-technical, manufactural and informational-management technologies, means and fractial measures during life cycle (LC) of high-quality manufacturing products (MP) for obtaining maximal required available level of quality and minimal product technical exploitational costs.

Contemporary standards being CALS vanguard methodology in not right measure answer necessary purposes. CALS standard have a debatable achievement and the following essential shortcoming:

- informational-technical-economic models being not dynamical;
- integrated database for analysis of logistic support is super plus on one hand and on the other hand does not contain information necessary for complex through cost LC estimation according to modern decision support algorithms;
- computational algorithms for various LC stage are simplified and do not permit forecasting with necessary accuracy and perform at conditions of internal and external noises and stochastic factors.

So ILS standard do not provide the whole realization of advantages for modern and perspective information technologies (IT) including staff structure in the field of stochastic modeling and estimation of two interconnected spheres: techno-sphere (techniques and technologies) and social ones.

These stochastic systems (StS) form the new systems class: OTES-CALS systems. Such systems destined for the production and realization of various services including engineering and other categorical works providing exploitation, aftersale MP support and repair, staff, medical, economical and financial support of all processes. New developed approach is based on new stochasting modeling and estimating approaches. Nowadays such IT are widely used in technical application of complex systems functioning in stochastic media.

Estimation of IT is based on: (1) model of OTES; (2) model of OTES-O (observation system); (3) model OTES-N (noise support); (4) criteria, estimation methods models and for new generations of synergetic OTES (SOTES) measuring model and organization-production block (OPB) in OTES-O are separated.

Synergetics being interdisciplinary science is based on the principle of self-realization of the open nonlinear dissipative and nonconservative systems. According to [6, 7] in equilibrium when all systems parameters are stable and variation in it arise due to minimal deviations of some control parameters. As a result, the system begins to move off from equilibrium state with increasing velocity. Further the non-stability process lead to total chaos and as a result appears bifurcation. After that gradually new regime establishes and so on.

The existence of big amount of free entering elements and subsystems of various levels is the basic principle of self-organization. One of inalienable properties of synergetical system is the existence of "attractors". Attractor is defined as attraction set (manifold) in phase space being the aim of all nonlinear trajectories of moving initial point (IP). These manifolds are time invariant and are defined from equilibrium equation. Invariant manifolds are also determined as constraints of non-conservative synergetical system. In synergetical control theory [8] transition from natural, unsupervised behavior according to algorithms of dissipative structure to control motion IP along artificially in putted demanded invariant manifolds. As a control object of synergetical system always nonlinear its dynamics may be described by nonlinear differential equations. In case of big dimension the parameters of order are introduced by revealing most slow variable and more quick subordination variables. This approach in hierarchical synergetic system is called subordination principle. So at lower hierarchy level processors go with maximal velocity. Invariant manifolds are connected with slow dynamics.

Section 1 is devoted to probabilistic modeling problems in typical StS. Special attention is paid to hybrid systems. Such specific StS as linear, linear with the Gaussian parametric noises and nonlinear reducible to quasilinear by normal approximation method. For quick off-line and on-line application theory of conditionally optimal forecasting in typical StS is developed in Section 2. In Section 3 basic off-line algorithm of probability modeling in SOTES are presented. Basic conditionary optimal filtering and forecasting quick – off-line and on-line algorithms for SOTES are given in Section 4. Peculiarities of new SOTES generalizations are described in Section 5. Simple example illustrating the influence of SOTES-N noise on rules and functional indexes of subsystems accompanying life cycle production, its filtration and forecasting is presented in Section 6. Experimental software tools for forecasting of cost and technical readiness for aircraft parks are developed.

2. Probabilistic modeling in StS

Let us consider basic mathematical models of stochastic OTES:

- continuous models defined by stochastic differential equations;
- discrete models defined by stochastic difference equations;
- hydride models as a mixer of difference and differential equations.

Probabilistic analytical modeling of stochastic systems (StS) equations is based on the solution of deterministic evolutionary equations (Fokker-Plank-Kolmogorov, Pugachev, Feller-Kolmogorov) for one- and finite dimensions. For stochastic equations of high dimensions solution of evolutionary equation meets principle computationary difficulties.

At practice taking into account specific properties of StS it is possible to design rather simple stochastic models using a priori data about StS structure, parameters and stochastic factors. It is very important to design for different stages of the life cycle (LC) models based on available information. At the last LC stage we need hybrid stochastic models.

Let us consider basic general and specific stochastic models and basic algorithms of probabilistic analytical modeling. Special attention will paid to algorithms based on normal approximation, statistical linearization and equivalent linearization methods. For principally nonlinear non Gaussian StS may be recommended corresponding parametrization methods [9].

2.1 Continuous StS

Continuous stochastic models of systems involve the action of various random factors. While using models described by differential equations the inclusion of random factors leads to the equations which contain random variables.

Differential equations for a StS (more precisely for a stochastic model of a system) must be replaced in the general case by the Equations [9, 10].

$$\dot{Z} = F(Z,x,t), \quad Y = G(Z,t), \tag{1}$$

where $F(z,x,t)$ and $G(z,t)$ are random functions of the p-dimensional vector, z, n-dimensional vector x and time t (as a rule G is independent of x). In consequence of the randomness of the right-hand sides of Eq. (1) and also perhaps of the initial value of the state vector $Z_0 = Z(t_0)$ the state vector of the system Z and the output Y represent the random variables at any fixed time moment t. This is the reason to denote them by capital letters as well as the random functions in the right-hand sides to the Eq. (1). The state vector of the system $Z(t)$ and its output $Y(t)$ considered as the functions of time t represent random functions of time t (in the general case vector random functions). In every specific trial the random functions $F(z,x,t)$ and $G(z,t)$ are realized in the form of some functions $f(z,x,t)$ and $g(z,t)$ and these realizations determine the corresponding realizations $z(t), y(t)$ of the state vector $Z(t)$ and the output $Y(t)$ satisfying the differential equations (which are the realizations of Eq. (1)

$$\dot{z} = f(z,x,t), \quad y = g(z,t).$$

Thus we come to the necessity to study the differential equations with random functions in the right-hand sides.

At practice the randomness of the right-hand sides of the differential equations arises usually from the fact that they represent known functions some of whose arguments are considered as random variables or as random functions of time t and perhaps of the state and the output of the system. But in the latter cased these functions are usually replaced by the random functions of time which are only obtained by assuming that their arguments Z and Y are known functions of time corresponding to the nominal regime of system functioning. In practical problems such an assumption usually provides sufficient accuracy.

So we may restrict ourselves to the case where all uncertain variables in the right-hand sides of differential equations may be considered as random functions of time. Then Eq. (1) may be written in the form

$$\dot{Z} = f(Z,x,N_1(t),t), \quad Y = g(Z,N_2(t),t), \tag{2}$$

where f and g are known functions whose arguments include random functions of time $N_1(t)$ and $N_2(t)$. The initial state vector of the system Z_0 in practical problems is always a random variable independent of the random functions $N_1(t)$ and $N_2(t)$ (independent of random disturbances acting of the system).

Every realization $\left[n_1(t)^T n_2(t)^T\right]^T$ of the random function $\left[N_1(t)^T N_2(t)^T\right]^T$ determines the corresponding realizations $f(z,x,n_1(t),t)$, $g(z,n_2(t),t)$ of the functions $f(z,x,N_1(t),t)$, $g(z,N_2(t),t)$, and in accordance with this Eq. (2) determine respective realizations $z(t)$ and $y(t)$ of the state vector of the system $Z(t)$ and its output $Y(t)$.

Following [9, 10] let us consider the differential equation

$$dX/dt = a(X,t) + b(X,t)V, \tag{3}$$

where $a(x,t)$, $b(x,t)$ being functions mapping $R^p \times R$ into R^p and R^{pq}, respectively, is called a stochastic differential equation if the random function (generalized) $V(t)$ represents a white noise in the strict sense. Let X_0 be a random vector of the same dimension as the random function $X(t)$. Eq. (3) with the initial condition $X(t_0) = X_0$ determines the stochastic process (StP)$X(t)$.

In order to give an exact sense to Eq. (3) and to the above statement we shall integrate formally Eq. (3) in the limits from t_0 to t with the initial condition $X(t_0) = X_0$. As result we obtain

$$X(t) = X_0 + \int_{t_0}^{t} a(X(\tau), \tau)d\tau + \int_{t_0}^{t} b(X(\tau), \tau)V(\tau)d\tau$$

where the first integral represents a mean square (m.s.) integral. Introducing the StP with independent increments $W(t)$ whose derivative is a white noise $V(t)$ we rewrite the previous equation in the form

$$X(t) = X_0 + \int_{t_0}^{t} a(X(\tau), \tau)d\tau + \int_{t_0}^{t} b(X(\tau), \tau)dW(\tau). \tag{4}$$

This equation has the exact sense. Stochastic differential Eq. (3) or the equivalent equation

$$dX = a(X, t)dt + b(X, t)dW \tag{5}$$

with the initial condition $X(t_0) = X_0$ represents a concise form for of Eq. (4).

Eq. (5) in which the second integral represents a stochastic Ito integral is called a stochastic Ito integral equation and the corresponding differential Eq. (3) or (5) is called a stochastic Ito differential Eq.

A random process $X(t)$ satisfying Eq. (4) in which the integral represent the m.s. limits of the corresponding integral sums is called a mean square of shortly, an m.s. solution of stochastic integral Eq. (4) and of the corresponding stochastic differential Eq. (3) or (5) with the initial condition $X(t_0) = X_0$.

If the integrals in Eq. (4) exist for every realization of the StP$W(t)$ and $X(t)$ and equality (4) is valid for every realization then the random process $X(t)$ is called a solution in the realization of Eq. (4) and of the corresponding stochastic differential Eq. (3) and (5) with the initial condition $X(t_0) = X_0$.

Stochastic Ito differential Eqs. (3) and (5) with the initial condition $X(t_0) = X_0$, where X_0 is a random variable independent of the future values of a white noise $V(s)$, $s > t_0$ (future increments $W(s) - W(t)$, $s > t \geq t_0$, of the process W) determines a Markov random process.

In case of W being vector StP with independent in increments probabilistic modeling of one and n-dimensional characteristic functions $g_1 = \mathrm{E}e^{i\lambda^T Z(t)}$ and $g_n = \mathrm{E}\exp\{i\sum_{k=1}^{n}\lambda_k^T Z(t_k)\}$ and densities f_1 and f_n is based on the following integrodifferential Pugachev Eqs:

$$\frac{\partial g_1(\lambda; t)}{\partial t} = \frac{1}{(2\pi)^p} \int_{-\infty}^{\infty} \int_{-\infty}^{\infty} \left[i\lambda^T a(z, t) + \chi\left(b(z, t)^T \lambda; t\right)\right] e^{i(\lambda^T + \mu^T)z} g_1(\mu; t) d\mu dz, \tag{6}$$

$$\begin{aligned} \frac{\partial}{\partial t_n} g_n(\lambda_1, \dots, \lambda_n; t_1, \dots, t_n) &= \frac{1}{(2\pi)^{pq}} \int_{-\infty}^{\infty} \int_{-\infty}^{\infty} \left[i\lambda_n^T a(z_n, t_n) + \chi\left(b(z_n, t_n)^T \lambda_n; t_n\right)\right] \\ &\times \exp\left\{i\sum_{k=1}^{n}(\lambda_k^T - \mu_k^T)z_K\right\} g_n(\mu_1, \dots, \mu_n; t_1, \dots, t_n) d\mu_1, \dots, d\mu_n; dz_1, \dots, dz_n, \end{aligned} \tag{7}$$

$$\frac{\partial f_1(z;t)}{\partial t} = \frac{1}{(2\pi)^p} \int_{-\infty}^{\infty} \int_{-\infty}^{\infty} \left[i\lambda^T a(\zeta,t) + \chi\left(b(\zeta,t)^T \lambda; t\right) \right] e^{i\lambda^T(\zeta - z)} f_1(\zeta,t) d\zeta d\lambda, \tag{8}$$

$$\frac{\partial}{\partial t_n} f_n(z_1, \dots, z_n; t_1, \dots, t_n) = \frac{1}{(2\pi)^{np}} \int_{-\infty}^{\infty} \dots \int_{-\infty}^{\infty} \left[i\lambda_n^T a(\zeta_n, t_n) + \chi\left(b(\zeta_n, t_n)^T \lambda_n; t_n\right) \right]$$
$$\times \exp\left\{ i \sum_{k=1}^{n} \lambda_k^T (\zeta_k - z_k) \right\} f_n(\zeta_1, \dots, \zeta_n; t_1, \dots, t_n) d\zeta_1, \dots, d\zeta_n; d\lambda_1, \dots, d\lambda_n, \tag{9}$$

$$f_1(z; t_0) = f_0(z), \tag{10}$$

where i being imaginary unit,

$$\chi(\mu; t) = \frac{1}{h_1(\mu; t)} \frac{\partial h_1(\mu; t)}{\partial t}, \tag{11}$$

$$f_n(z_1, \dots, z_{n-1}, z_n; t_1, \dots, t_{n-1}, t_{n-1}) = f_{n-1}(z_1, \dots, z_{n-1}; t_1, \dots, t_{n-1}) \delta(z_n - z_{n-1}). \tag{12}$$

For the Wiener W StP with intensity matrix $v(t)$ we use Fokker-Plank-Kolmogorov Eqs:

$$\frac{\partial f_n(z;t)}{\partial t} = \frac{\partial^T}{\partial z} \left[a(z,t) f_n(z;t) \right] + \frac{1}{2} \mathrm{tr} \left\{ \frac{\partial}{\partial z} \frac{\partial^T}{\partial z} b(z,t) v(t) b(z,t)^T f_n(z;t) \right\} \tag{13}$$

at initial conditions (12).

2.2 Discrete StS

For discrete vector StP yielding regression and autoregression StS

$$X_{k+1} = \omega_k(X_k, V_k) \quad (k = 1, 2, \dots), \tag{14}$$

$$X_{k+1} = a_k(X_k) + b_k(X_k) V_k \quad (k = 1, 2, \dots). \tag{15}$$

Eqs for one and n dimensional densities and characteristic functions are described by:

$$f_k(x) = \frac{1}{(2\pi)^p} \int_{-\infty}^{\infty} e^{-i\lambda^T x} g_k(\lambda) d\lambda, \quad g_k(\lambda) = \mathrm{E} \exp\left\{ i\lambda^T X_k \right\}, \tag{16}$$

$$f_{k_1, \dots, k_n}(x_1, \dots, x_n) = \frac{1}{(2\pi)^{np}} \int_{-\infty}^{\infty} \exp\left\{ i \sum_{h=1}^{n} \lambda_h^T x_h \right\} g_{k_1, \dots, k_n}(\lambda_1, \dots, \lambda_n) d\lambda_1, \dots, d\lambda_n, \tag{17}$$

$$g_{k_1, \dots, k_n}(\lambda_1, \dots, \lambda_n) = \mathrm{E} \exp\left\{ i \sum_{l=1}^{n} \lambda_l^T x_{k_l} \right\}, \tag{18}$$

$$g_{k+1}(\lambda) = \mathrm{E} \exp\left\{ i\lambda \omega_k(X_k, V_k) = \int_{-\infty}^{\infty} \int_{-\infty}^{\infty} e^{-i\lambda^T \omega_{\rightarrow}(x,v)} f_k(x) h_k(v) dx dv, \right. \tag{19}$$

$$g_{k_1,\dots,k_n}(\lambda_1,\dots,\lambda_n) = \mathrm{E}\exp\left\{i\sum_{l=1}^{n}\lambda_l^T x_{k_l} + i\lambda_n^T \omega_{k_n}(X_{k_n}, V_{k_n})\right\}$$
$$= \int_{-\infty}^{\infty}\dots\int_{-\infty}^{\infty}\int_{-\infty}^{\infty}\exp\left\{i\sum_{h=1}^{n}\lambda_h^T x_h + i\lambda_n^T \omega_{k_n}(x_n, v_n)\right\} f_{k_1,\dots,k_n}(x_1,\dots,x_n)\eta_n(v_n)dx_1\dots dx_n, dv_n. \quad (20)$$

Here E being symbol of mathematical expectation, h_k being V_k characteristic function

$$\begin{aligned} g_{k_1,\dots,k_{n-1},k_{n-1}}(\lambda_1,\dots,\lambda_n) &= g_{k_1,\dots,k_{n-1}}(\lambda_1,\dots,\lambda_{n-1}+\lambda_n), \\ g_{k_1,\dots,k_n}(\lambda_1,\dots,\lambda_n) &= g_{s_1,\dots,s_n}(\lambda_{s_1},\dots,\lambda_{s_n}), \end{aligned} \quad (21)$$

where $(s_1, \dots, s_n)$ – permutation of $(1, \dots, n)$ at $k_{s_1} < k_{s_2} < \dots < k_{s_n}$.

In case of the autoregression StS (1.14) basic characteristic functions are given by Eqs:

$$g_{k+1}(\lambda_1,\dots,\lambda_n) = \mathrm{E}\exp\left\{i\lambda^T a_k(x_k) + i\lambda^T b_k(X_k)V_k\right\}$$
$$= \int_{-\infty}^{\infty}\int_{-\infty}^{\infty} e^{i\lambda^T a_k(x) + i\lambda^T a_k(x)v} f_k(x)h_k(v)dxdv = \mathrm{E}\Big[\exp\left\{i\lambda^T a_k(X_k)\right\} + h_k\Big(b_k(X_k)^T\lambda\Big)\Big], \quad (22)$$

$$g_{k_1,\dots,k_n}(\lambda_1,\dots,\lambda_n) = \mathrm{E}\exp\left\{i\sum_{l=1}^{n-1}\lambda_l^T x_{k_l} + i\lambda_n^T a_{k_n}(X_{k_n}) + i\lambda_n^T b_{k_n}(X_{k_n})V_{k_n}\right\}$$
$$= \int_{-\infty}^{\infty}\dots\int_{-\infty}^{\infty}\int_{-\infty}^{\infty}\exp\left\{i\sum_{h=1}^{n-1}\lambda_l^T x_l + i\lambda_n^T a_{k_n}(x_n) + i\lambda_n^T b_{k_n}(x_n)v_n\right\} \quad (23)$$
$$\times \mathrm{E}\left[\exp\left\{i\sum_{l=1}^{n-1}\lambda_l^T x_{k_l} + i\lambda_n^T a_{k_n}(X_{k_n})\right\} + h_n\Big(b_n(X_n)^T\lambda_n\Big)\right].$$

2.3 Hybrid continuous and discrete StS

When the system described by Eq. (2) is automatically controlled the function which determines the goal of control is measured with random errors and the control system components forming the required input x^* are always subject to noises, i.e. to random disturbances. Forming the required input and the real input including the additional variables necessary to transform these equations into a first-order equation may be written in the form

$$\dot{X} = \varphi(X, U, t), \quad \dot{U} = \psi(X, Z, U, N_3(t), t) \quad (24)$$

where U is the vector composed of the required input and all the auxiliary variables, and $N_3(t)$ is some random function of time t (in the general case representing a vector random function). Writing down these equations we have taken into account that owing to the action of noises described by the random function $N_3(t)$, the vector U and the input X represent random functions of time and in accordance with this we denoted them by capital letters. These equations together with the first Eq. (2) form the set of Eqs

$$\dot{Z} = f(Z, X, N_1(t), t), \quad \dot{X} = \varphi(X, U, t), \quad \dot{U} = \psi(X, Z, U, N_3(t), t).$$

These equations may be written in the form of one equation determining the extended state vector of the system $Z_1 = \left[Z^T X^T U^T\right]^T$:

$$\dot{Z}_1 = f_1(Z_1, N_4(t), t)$$

where $N_4(t) = \left[N_1(t)^T N_3(t)^T\right]^T$, and

$$f_1(Z_1, N_4(t), t) = \left[f(Z, X, N_1, t)^T \ \varphi(X, U, t)^T \ \psi(X, Z, U, N_3, t)^T\right]^T.$$

As a result rejecting the indices of Z_1 and f_1 we replace the set of the Eqs. (2) and (24) by the equations

$$\dot{Z} = f_1(Z_1, N_4(t), t), \quad Y = g(Z, N_2(t), t).$$

In practical problems the random functions $N_1(t)$ and $N_2(t)$ are practically always independent. But the random function $N_3(t)$ depends on $N_1(t)$ and $N_2(t)$ due to the fact that the function $h(Y, t) = h(g(Z, N_2(t), t), t)$ and its total derivative with respect to time t enter into Eq. (24). Therefore, the random function $N_2(t)$ and $N_4(t)$ are dependent. Introducing the composite vector random function $N(t) = \left[N_1(t)^T N_2(t)^T N_3(t)^T\right]^T$ we rewrite the equations obtained in the form

$$\dot{Z} = f_1(Z_1, N(t), t), \quad Y = g(Z, N(t), t). \tag{25}$$

Thus in the cased of an automatically controlled system described by Eq. (2), after coupling Eq. (2) with the equations of forming the required and the real inputs we come to the equations of the form of (23) containing the random function $N(t)$.

If a control StS based on digital computers we decompose the extended state vector Z into two subvectors Z', Z'', $Z = \left[Z'^T Z''^T\right]^T$ one of which Z' represents a continuously varying random function, and the other Z'' is a step random function varying by jumps at prescribed time moments $t^{(k)}$ $(k = 0, 1, 2, \ldots)$. Then introducing the random function

$$Z''(t) = \sum_{k=0}^{\infty} Z''_k \mathbf{1}_{A_k}(t)$$

and putting $Z'_k = Z'\left(t^{(k)}\right)$ $(k = 0, 1, 2, \ldots)$ we get the set of equations describing the evaluation of the extended state vector of controlled

$$\dot{Z} = f(Z, N(t), t), \quad Z''_{k+1} = \varphi_k(Z_k, N_k) \tag{26}$$

where N_k $(k = 0, 1, 2, \ldots)$ are some random variables, and $N(t)$ some random function.

For hybrid StS (HStS) let us now consider the case of a discrete-continuous system whose state vector $Z = \left[Z'^T Z''^T\right]^T$ (extended in the general case) is determined by the set of equations

$$\dot{Z}' = a(Z,t) + b(Z,t)V, \quad Z'' = \sum_{k=0}^{\infty} Z''_k \mathbf{1}_{A_k}(t), \quad Z''_{k+1} = \omega_k(Z_k, V_k) \tag{27}$$

where is the value of $Z(t)$ at $t = t^{(k)}$, $Z_k = \left[Z'^T_k Z''^T_k\right]^T = Z\left(t^{(k)}\right)(k = 0, 1, 2, \dots)$, a, b, ω_k are functions of the arguments indicated $\mathbf{1}_{A_k}(t)$ is the indicator of the interval $A_k = \left[t^{(k)} t^{(k+1)}\right)$ $(k = 0, 1, 2, \dots)$, V is a white noise in the strict sense, $\{V_k\}$ is a sequence of independent random variables independent of the white noise V. The one-dimensional characteristic function $h_1(\mu; t)$ of the process with independent increments $W(t)$ whose weak m.s. derivative is the white noise V, and the distributions of the random variables V_k will be assumed known.

Introducing the random processes

$$Z''(t) = \sum_{k=0}^{\infty} Z'_k \mathbf{1}_{A_k}(t), \quad \overline{Z}(t) = \left[Z'(t)^T Z''(t)^T Z'''(t)^T\right]^T.$$

we derive in the same way as before the equation for the one-dimensional characteristic function

$$\begin{aligned} g_1(\lambda; t) = \mathrm{E} e^{i\lambda^T \overline{Z}(t)} &= \mathrm{E} \exp\left\{ i\lambda'^T Z'(t) + i\lambda''^T Z''(t) + i\lambda'''^T Z'''(t) \right\} \\ &= \mathrm{E} \exp\left\{ i\lambda'^T Z'(t) + i\lambda''^T Z''_k + i\lambda'''^T Z'_k \right\} \end{aligned}$$

of the StP $\overline{Z}(t)$

$$\frac{\partial g_1(\lambda; t)}{\partial t} = \mathrm{E}\left\{ i\lambda^T a(Z,t) + \chi(b(Z,t)^T \lambda^T; t) e^{i\lambda^T \overline{Z}} \right\}. \tag{28}$$

Taking the initial moment $t_0 = t^{(0)}$ the initial condition for Eq. (25) is

$$g_1(\lambda; t_0) = \mathrm{E}\left\{ \left(i\lambda'^T + i\lambda'''^T \right) Z'_0 + i\lambda''^T Z''_0 \right\} = g_0\left(\left[\lambda'^T + \lambda'''^T \quad \lambda''^T \right]^T \right) \tag{29}$$

where $g_0(\rho)$ is the characteristic function of the initial value $Z_0 = Z(t_0)$ of the process $Z(t)$.

At the moment $t^{(k)}$ the value of $g_1(\lambda; t)$ is evidently equal to

$$\mathrm{E} \exp\left\{ i\left(\lambda'^T + \lambda'''^T \right) Z'_k + i\lambda''^T Z''_k \right\},$$

i.e. to the value $g_k\left(\left[\lambda'^T + \lambda'''^T \quad \lambda''^T \right]^T \right)$ of the characteristic function $g_k(\rho)$ of the random variable $Z_k = \left[Z''^T_k \ Z''^T_k \right]^T$. If the function $\chi(\mu; t)$ is continuous function of t at any μ the $g_1(\lambda; t)$ tends to

$$\mathrm{E}\left\{i\lambda'^T Z'_{k+1} + i\lambda''^T Z''_k + i\lambda'''^T Z'_k\right\}$$

when $t \to t^{(k+1)}$, i.e. to the joint characteristic function $g'_k(\lambda', \lambda'', \lambda''')$ of the random variables $Z'_{k+1}, Z''_k.Z'_k$

$$g_1\left(\lambda; t^{(k+1)} - 0\right) = \lim_{t \to t^{(k+1)}} g_1(\lambda; t) = g'_k(\lambda', \lambda'', \lambda''').$$

At the moment $t^{(k+1)} g_1(\lambda; t)$ changes its value by the jump and becomes equal to

$$\mathrm{E}\ \exp\left\{i\left(\lambda'^T + \lambda'''^T\right)Z'_{k+1} + i\lambda''^T Z''_{k+1}\right\} = g_{k+1}\left(\left[\lambda'^T + \lambda'''^T \quad \lambda''^T\right]^T\right).$$

To evaluate this, we substitute here the expression of Z''_{k+1} from the last Eq. (27). Then we get

$$g_1\left(\lambda; t^{(k+1)}\right) = \mathrm{E}\ \ \exp\left\{i\left(\lambda'^T + \lambda'''^T\right)Z'_{k+1} + i\lambda''^T \omega_k(Z_k, V_k)\right\}. \tag{30}$$

Owing to the independence of the sequence of random variables $\{V_k\}$ of the white noise V and independence of V_k of $V_0, V_1, \ldots, V_{k-1}$ the random variables Z_k and Z'_{k+1} are independent of V_k. Hence, the expectation in the right-hand side of Eq. (30) is completely determined by the known distribution of the random variable V_k and by the joint characteristic function $g'_k(\lambda', \lambda'', \lambda''')$ of the random variables Z'_{k+1}, Z''_k, Z'_k, i.e. by $g_1(\lambda; t^{(k+1)} - 0)$. So Eq. (26) with the initial condition (27) and formula (28) determine the evolution of the one-dimensional characteristic function $g_1(\lambda; t)$ of the process $\overline{Z}(t) = \left[Z'(t)^T Z''(t)^T Z'''(t)^T\right]^T$ and its jump-wise increments at the moments $t^{(k)}$ $(k = 1, 2, \ldots)$.

In the case of the discrete-continuous HStS whose state vector is determined by Eqs

$$\dot{Z} = a(Z, t) + b(Z, t)V \tag{31}$$

we get in the same way the equation for the n-dimensional characteristic function $g_n(\lambda_1, \ldots, \lambda_n; t_1, \ldots, t_n)$ of the random process $Z(t) = \left[Z'(t)^T Z''(t)^T Z'''(t)^T\right]^T$,

$$\begin{aligned} &\partial g_n(\lambda_1, \ldots, \lambda_n; t_1, \ldots, t_n)/\partial t_n \\ &= \mathrm{E}\left[i\lambda_n^T a(Z(t_n), t_n) + \chi\left(b(Z(t_n), t_n)^T \lambda_n; t_n\right)\right]\{i\lambda_1^T \overline{Z}(t_1) + \cdots + i\lambda_n^T \overline{Z}(t_n)\}, \end{aligned} \tag{32}$$

And the formula for the value of $g_n(\lambda_1, \ldots, \lambda_n; t_1, \ldots, t_n)$ at $t_n = t^{(k+1)} \geq t_{n-1}$,

$$\begin{aligned} g_n\left(\lambda_1, \ldots, \lambda_n; t_1, \ldots, t_{n-1}, t^{(k+1)}\right) = \mathrm{E}\Big\{ & i\lambda_1^T \overline{Z}(t_1) + \cdots + i\lambda_{n-1}^T \overline{Z}(t_{n-1}) \\ & + i(\lambda_n'^T + \lambda_n'''^T) Z'_{k+1} + i\lambda_n''^T \omega_k(Z_k, V_k)\Big\}. \end{aligned} \tag{33}$$

At the point $t_n = t^{(k+1)} g_n(\lambda_1, \dots, \lambda_n; t_1, \dots, t_n)$ changes its value by jump from

$$g_n\left(\lambda_1, \dots, \lambda_n; t_1, \dots, t_{n-1}, t^{(k+1)} - 0\right) = \mathrm{E}\left\{ i\lambda_1^T \overline{Z}(t_1) + \dots + i\lambda_{n-1}^T \overline{Z}(t_{n-1}) + i\lambda_n'^T Z'_{k+1} + i\lambda_n''^T Z''_k + i\lambda_n'''^T Z'_k \right\}$$

to $g_n\left(\lambda_1, \dots, \lambda_n; t_1, \dots, t_{n-1}, t^{(k+1)}\right)$ given by (33).

The right-hand side of (33) is completely determined by the known distribution of the random variable V_k and by the joint characteristic function $g_n\left(\lambda_1, \dots, \lambda_n; t_1, \dots, t_{n-1}, t^{(k+1)} - 0\right)$ of the random variables $Z(t_1), \dots, \overline{Z}(t_{n-1}), Z'_{k+1}, Z''_k, Z'_k$. Hence, Eq. (32) with the corresponding initial condition and formula (33) determine the evolution and the jump-wise increments of $g_n(\lambda_1, \dots, \lambda_n; t_1, \dots, t_n)$ at the points $t^{(k+1)}$ when t_n increases starting from the value t_{n-1}.

2.4 Linear StS

For differential linear StS and W being StP with independent increments $V = \dot{W}$

$$\dot{Z} = aZ + a_0 + bV \tag{34}$$

corresponding Eqs for n-dimensional characteristic function are as follows:

$$\frac{\partial g_n}{\partial t_n} = \lambda_n^T a(t_n) \frac{\partial g_n}{\partial \lambda_n} + \left[i\lambda_n^T a_0(t_n) + \chi\left(b(t_n)^T \lambda_n; t_n\right) \right] g_n. \tag{35}$$

Explit formulae for n-dimensional characteristic function is described by formulae

$$\begin{aligned} g_n(\lambda_1, \dots, \lambda_n; t_1, \dots, t_n) = {} & g_0\left(\sum_{k=1}^n u(t_k, t_0)^T \lambda_k \right) \exp\left\{ i \sum_{k=1}^n \lambda_k^T \int_{t_0}^{t_k} u(t_k, \tau)^T a_0(\tau) d\tau \right. \\ & \left. + \sum_{k=1}^n \int_{t_{k-1}}^{t_k} \chi\left(b(\tau)^T \sum_{l=k}^n u(t_1, \tau)^T \lambda_l; \tau \right) d\tau \right\} \quad (n = 1, 2, \dots). \end{aligned} \tag{36}$$

Here $u = u(t_k, \tau)$ being fundamental solution of Eq $\dot{u} = au$ at condition: $u(t, t) = I$ (unit $(n \times n)$ matrix).

In case of the Gaussian white noise V with intensity matrix v characteristic function g_n is Gaussian

$$\begin{aligned} g_n(\lambda_1, \dots, \lambda_n; t_1, \dots, t_n) = {} & g_0\left(\sum_{k=1}^n u(t_k, t_0)^T \lambda_k \right) \exp\left\{ i \sum_{k=1}^n \lambda_k^T \int_{t_0}^{t_k} u(t_k, \tau) a_0(\tau) d\tau \right. \\ & \left. - \frac{1}{2} \sum_{l;h=1}^n \lambda_l^T \int_{t_0}^{\min(t_l, t_h)} u(t_l, \tau) b(\tau) v(\tau) b(\tau)^T u(t_h, \tau)^T d\tau \lambda_h \right\} \quad (n = 1, 2, \dots). \end{aligned} \tag{37}$$

2.5 Linear StS with the parametric Gaussian noises

In the case of StS with the Gaussian discrete additive and parametric noises described by Eq

$$\dot{Z} = aZ + a_0 + \left(b_0 + \sum_{h=1}^{p} b_h Z_h\right)V. \tag{38}$$

we have the infinite set of equations which in this case is decomposed into independent sets of equations for the initial moments α_k of each given order

$$\dot{\alpha}_k = \sum_{r=1}^{p} k_r\left(a_{r,0}\alpha_{k-e_r} + \sum_{q=1}^{p} a_{r,e_q}\alpha_{k+e_q-e_r}\right)$$
$$+\frac{1}{2}\sum_{r=1}^{p} k_r(k_r-1)\left(\sigma_{rr,0}\alpha_{k-2e_r} + \sum_{q=1}^{p}\sigma_{rr,e_q}\alpha_{k+e_q-2e_r} + \sum_{q,u=1}^{p}\sigma_{rr,e_q+e_u}\alpha_{k+e_q+e_u-2e_r}\right)$$
$$+\sum_{r=2}^{p}\sum_{s=1}^{p-1} k_r k_s\left(\sigma_{rs,0}\alpha_{k-e_r-e_s} + \sum_{q=1}^{p}\sigma_{rs,e_q}\alpha_{k+e_q-e_r-e_s} + \sum_{q,u=1}^{p}\sigma_{rs,e_q+e_u}\alpha_{k+e_q+e_u-e_r-e_s}\right), \tag{39}$$

$$a_{r,0} = a_{0,r}, \quad a_{r,e_q} = a_{rq} \quad (k_1, \dots, k_p = 0, 1, 2, \dots; \quad \sigma(k) = 1, 2, \dots).$$

Corresponding Eqs of correlational theory are as follows:

$$\dot{m} = am + a_0; \tag{40}$$

$$\dot{K} = aK + Ka^T + b_0 v b_o^T + \sum_{h=1}^{p}(b_h v b_0^T + b_0 v b_h^T)m_0 + \sum_{h,l=1}^{p} b_h v b_l^T(m_h m_l + k_{hl}). \tag{41}$$

where k_{hl} is the covariance of the components Z_h and Z_l of the vector $Z \quad (h, l = 1, \dots, p)$. Eq. (41) with the initial condition $K(t_0) = K_o\left(k_{pq}(t_0) = k_{pq}^0\right)$ completely determines the covariance matrix $K(t)$ of the vector $Z(t)$ at any time moment t after funding its expectation m.

For discrete StS with the Gaussian parametric noises correlational Eqs may be presented in the following form:

$$X_{k+1} = a_k X_k + a_{0l} + \left(b_{0l} + \sum_{j=1}^{p} b_{kj}X_{kj}\right)V_k, \tag{42}$$

$$m_{k+1} = a_k m_k + a_{0k}, \quad m_k = \mathrm{E}Y_k, \tag{43}$$

$$K_{k+1} = a_k K_k a_k^T + b_{0k}v_k b_{0k}^T + \sum_{j=1}^{p}\left(b_{0k}v_l b_{jk}^T + b_{jk}v_l b_{0k}^T\right)m_{jk}$$
$$+ \sum_{j=1}^{p}\sum_{h=1}^{p} b_{jk}v_k b_{hk}^T(m_{kj}m_{kh} + k_{kjh}), \tag{44}$$
$$K_1 = E(Y_1 - m_1)(Y_1 - m_1)^T,$$

$$K(j,h+1) = K(j,h)a_h^T, \quad K(j,j) = K. \tag{45}$$

2.6 Normal approximation method

For StS of high dimensions methods of normal approximation (MNA) are the only used at engineering practice. In case of additive noises $b(x,t) = b_0(t)$ MNA is known as the method of statistical linearization (MSL).

Basic Eqs of MNA are as follows [9]:

$$\begin{gathered} g_1(\lambda;t) \approx \exp\left\{i\lambda^T m_t - \frac{1}{2}\lambda^T K_t \lambda\right\}, \\ f_1(x;t) \approx \left[(2\pi)^p |K_t|\right]^{-1/2} \exp\left\{-\frac{1}{2}(x^T - m_t^T)K_t^{-1}(x - m_t)\right\}, \end{gathered} \tag{46}$$

$$\dot{m}_t = \varphi_1(m_t, K_t, t) \quad m(t_0) = m_0, \varphi_1(m_t, K_t, t) = \mathrm{E}_{\mathcal{N}} a(Y_t, t), \tag{47}$$

$$\dot{K}_t = \varphi_2(m_t, K_t, t) \quad K(t_0) = K_0, \tag{48}$$

$$\begin{gathered} \varphi_2(m_t, K_t, t) = \varphi_{21}(m_t, K_t, t) + \varphi_{21}(m_t, K_t, t)^T + \varphi_{22}(m_t, K_t, t), \\ \varphi_{21}(m_t, K_t, t) = \mathrm{E}_{\mathcal{N}} a(X_t, t)\left(X_t^T - m_t^T\right), \varphi_{22}(m_t, K_t, t) = \mathrm{E}_{\mathcal{N}} b(X_t, t) v(t) b(X_t, t)^T, \\ \frac{\partial K(t_1, t_2)}{\partial t_2} = K(t_1, t_2) K(t_2)^{-1} \varphi_{21}\left(m, (t_2) K(t_2), (t_2)^T,\right. \end{gathered} \tag{49}$$

$$\begin{gathered} g_n(\lambda_1, \dots, \lambda_n; t_1, \dots, t_n) = \exp\left\{i\lambda^T \overline{m}_n - \frac{1}{2}\overline{\lambda}^T \overline{K}_n \overline{\lambda}\right\} \quad (n = 1, 2, \dots), \\ f_n(x_1, \dots, x_n; t_1, \dots, t_n) = \left[(2\pi)^n \left|\overline{K}_n\right|^{-1/2} \exp\left\{\frac{1}{2}(\overline{x}_n^T - \overline{m}_n^T)\overline{K}_n^{-1}(\overline{x}_n - \overline{m}_n)\right\} \quad (n = 1, 2, \dots),\right. \end{gathered} \tag{50}$$

$$\begin{gathered} \overline{\lambda} = \left[\lambda_1^T \lambda_2^T \dots \lambda_n^T\right]^T, \quad \overline{m}_n = \left[m_x(t_1)^T m_x(t_2)^T \dots m_x(t_n)^T\right]^T, \\ \overline{K}_n = \begin{bmatrix} K(t_1,t_1) & K(t_1,t_2) & \dots & K(t_2,t_n) \\ K(t_2,t_1) & K(t_2,t_2) & \dots & K(t_2,t_n) \\ \vdots & \vdots & \vdots & \vdots \\ K(t_n,t_1) & K(t_n,t_2) & \dots & K(t_n,t_n) \end{bmatrix}, \text{where} \quad \overline{x}_n = \left[x_1^T x_2^T \dots x_n^T\right]^T. \end{gathered} \tag{51}$$

Eq. (49) may be rewritten in form

$$\frac{\partial K(t_1, t_2)}{\partial t_2} = \varphi_3(K(t_1, t_2), t_1, t_2) \tag{52}$$

where

$$\begin{gathered} \varphi_3(K(t_1, t_2), t_1, t_2) = \left[(2\pi)^{2n_x} \left|\overline{K}_2\right|^{-1/2} \int_{-\infty}^{\infty}\int_{-\infty}^{\infty} (x_1 - m_{t_1})\varphi(x_2, t_2) \right. \\ \times \exp\left\{-\left(x_1^T x_2^T\right) - \overline{m}_2^T)\overline{K}_2^{-1}\left(x_1^T x_2^T\right) - \overline{m}_2^T)\right\} dx_1 dx_2; \\ \overline{m}_2 = \left[m_{t_1}^T m_{t_2}^T\right]^T; \quad \overline{K}_2 = \begin{bmatrix} K(t_1,t_1) & K(t_1,t_2) \\ K(t_2,t_1) & K(t_2,t_2) \end{bmatrix}. \end{gathered} \tag{53}$$

For discrete StS equations of MNA may be presented in the following form:

$$m_{l+1} = \mathrm{E}_{\mathcal{N}}\omega_l(X_l, V_l) \quad m_1 = \mathrm{E}X_1 \quad (l = 1, 2), \tag{54}$$

$$K_{l+1} = \mathrm{E}_{\mathcal{N}}\omega_l(X_l, V_l)\omega_l(X_l, V_l)^T - \mathrm{E}_{\mathcal{N}}\omega_l(X_l, V_l)\mathrm{E}_{\mathcal{N}}\omega_l(X_l, V_l)^T, \tag{55}$$

at conditions

$$\begin{aligned} &K_1 = \mathrm{E}_{\mathcal{N}}(X_1 - m_1)(X_1 - m_1)^T (l = 1, 2), \\ &K_{lh} = \mathrm{E}_{\mathcal{N}} X_l \omega_h(X_h, V_h)^T - m_l \mathrm{E}_{\mathcal{N}}\omega_h(X_h, V_h)^T, \\ &K_{ll} = K_l \quad at \quad l < h (h = 1, 2, \ldots), \quad K_{\mathrm{ln}} = K(h, l) = K(h, l)^T at \quad l < h. \end{aligned} \tag{56}$$

Corresponding MNA equations for Eq. (15) are the special case of Eqs. (54)–(56).

3. Conditionally optimal forecasting in StS

Optimal forecasting is well developed for linear StS and off-line regimes [9]. For nonlinear StS linear StS with the parametric Gaussian noises and on-line regimes different versions approximate (suboptimal) methods are proposed. In [9] general results for complex statistical criteria and Bayes criteria are developed. Let us consider m.s. conditionally optimal forecasters for StS being models of stochastic OTES.

3.1 Continuous StS

Conditionally optimal forecasting (COFc) for mean square error (mse) criterion was suggested by Pugachev [10]. Following [9] we define COFC as a forecaster from class of admissible forecasters which at any joint distributions of variables X_t (state variable) $\hat{X}_t$ (estimate of X_t), Y_t (observation variable) at forecasting time $\Delta > 0$ and time moments $t \geq t_0$ in continuous (differential) StS

$$dX_t = a(X_t, Y_t, t)dt + b(X_t, Y_t, t)dW_1, \quad dY_t = a_1(X_t, Y_t, t)dt + b_1(X_t, Y_t, t)dW_2 \tag{57}$$

(W_1, W_2 being independent white noises with the independent increments; $\varphi\varphi_1\psi\psi_1$ being known nonlinear functions) gives the best estimate of $X_{s+\Delta}$ at infinitesimal time moment $s > t, \ s \to t$ realizing minimum $\mathrm{E}\left|\hat{X}_s - \hat{X}_{s+\Delta}\right|^2$. Then COFc at any time moment $t \geq t_0$ is reduced to finding optimal coefficients $\alpha_t, \beta_t, \gamma_t$ in the following Eq:

$$d\hat{X}_t = \alpha_t \xi\left(\hat{X}_t, Y_t, t\right)dt + \beta_t \eta\left(\hat{X}_t, Y_t, t\right)dY_t + \gamma_t dt. \tag{58}$$

Here $\xi = \xi\left(\hat{X}_t, Y_t, t\right)$, $\eta = \eta\left(\hat{X}_t, Y_t, t\right)$ are given functions of current observations Y_t, estimate $\hat{X}_t$ and time t.

Using theory of conditionary optimal estimation (13, 17, 18) for Eq

$$dX_{t+\Delta} = a(X_{t+\Delta}, t + \Delta)dt + b(X_{t+\Delta}, t + \Delta)dW_1(t + \Delta). \tag{59}$$

we get the following Eqs for coefficients $\alpha_t, \beta_t, \gamma_t$

$$\alpha_t m_1 + \beta_t m_2 + \gamma_t = m_0, \quad m_0 = \mathrm{E}a(X_t, Y_t, t), \quad m_1 = \mathrm{E}\xi\big(Y_t, \hat{X}_{t+\Delta}, t\big), \\ m_2 = \mathrm{E}\eta\big(Y_t, \hat{X}_{t+\Delta}, t\big)a_1(X_t, Y_t, t), \tag{60}$$

$$\beta_t = \kappa_{02}\kappa_{22}^{-1}, \quad \kappa_{02} = \mathrm{E}\big(X_t - \hat{X}_{t+\Delta}\big)a_1(X_t, Y_t, t)^T\eta\big(Y_t, \hat{X}_{t+\Delta}, t\big)^T \\ +\mathrm{E}b(X_t, Y_t, t)v(t)b_1(X_t, Y_t, t)^T\eta\big(Y_t, \hat{X}_{t+\Delta}, t\big)^T, \tag{61}$$

$$\kappa_{22} = \mathrm{E}\big(Y_t, \hat{X}_{t+\Delta}, t\big)b_1(X_t, Y_t, t)v(t)b_1(X_t, Y_t, t)^T\eta\big(Y_t, \hat{X}_{t+\Delta}, t\big)^T \\ +\mathrm{E}b(X_t, Y_t, t)v(t)b_1(X_t, Y_t, t)^T\eta\big(Y_t, \hat{X}_{t+\Delta}, t\big)^T \tag{62}$$

at condition $(\det \kappa_{22} \neq 0)$.

The theory of conditionally optimal forecasting gives the opportunity for simultaneous filtering of state and identification of StS parameters for different forecasting time Δ. All complex calculations for COFc design do not need current observations and may be performed on a priori data during design procedures. Practical application of such COFc is reduced to Eq. (58) integration. The time derivative for the error covariance matrix R_t is defined by formulae

$$\dot{R}_t = \mathrm{E}\Bigg[\big(X_{t+\Delta} - \hat{X}_t\big)a(X_{t+\Delta}, t+\Delta)^T + a(X_{t+\Delta}, t+\Delta)\Big(X_{t+\Delta}^T - \hat{X}_t^T\Big) \\ - \beta_t\eta\big(Y_t, \hat{X}_{t+\Delta}, t\big)b_1(X_t, Y_t, t)v_2(t)b_1(X_t, Y_t, t)^T\eta\big(Y_t, \hat{X}_{t+\Delta}, t\big)^T\beta_t^T \\ + b(X_{t+\Delta}, t+\Delta)v_1(t+\Delta)b\big(\hat{X}_{t+\Delta}, t+\Delta\big)^T\Bigg]. \tag{63}$$

Mathematical expectations in Eq. (60)–(63) are computed on the basis of joint distribution of random variables $\Big[X_t^T X_{t+\Delta}^T, Y_t^T, \hat{X}_t^T \hat{X}_{t+\Delta}^T\Big]^T$ by solution of the following Pugachev Eq for characteristic function $g_2(\lambda_1, \lambda_2, \lambda_3, \mu_1, \mu_2, \mu_3; t, s)$ for StP $\Big[X_t^T Y_t^T \hat{X}_t^T\Big]^T$ at $s > t$:

$$\partial g_2(\lambda_1, \lambda_2, \lambda_3, \mu_1, \mu_2, \mu_3; t, s)/\partial s = \mathrm{E}\Bigg\{i\mu_1^T a_1(Y_s, X_s, s) + i\mu_2^T a(X_s, s) \\ + i\mu_3^T\big[\alpha_s\xi\big(Y_s, \hat{X}_s, s\big) + \beta_s\eta(Y_s, X_s, s) + \gamma_s\big] \\ + \chi\Bigg(b_1(Y_s, X_s, s)^T\mu_1 + b(X_s, s)\mu_2 \\ + b_1(Y_s, X_s, s)^T\eta\big(Y_s, \hat{X}_s, s\big)^T\beta_s^T\mu_3; s\Bigg)\Bigg\} \\ \times \exp\Bigg\{i\lambda_1^T Y_t + i\lambda_2^T X_t + i\lambda_3^T\hat{X}_t + i\mu_1^T Y_s + i\mu_2^T X_s + i\mu_3^T\hat{X}_s\Bigg\}. \tag{64}$$

at condition

$$g_2(\lambda_1, \lambda_2, \lambda_3, \mu_1, \mu_2, \mu_3; t, t) = g_1(\lambda_1 + \mu_1, \lambda_2 + \mu_2, \lambda_3 + \mu_3; t, s). \tag{65}$$

Basic algorithms are defined by the following Proposals .3.1.1–3.1.3.

Proposal 3.1.1. *At the conditions of the existence of probability moments (60), (61) nonlinear COFc is defined by* Eqs. (58) and, (63).

Proposal 3.1.2. *For linear differential StS*

$$dX_t = (a_1 X_t + a_0)dt + bdW_1, \quad dY_t = (bY_t + b_1 X_t + b_0)dt + b_1 dW_2. \tag{66}$$

Eqs of exact COFc are as follows:

$$d\hat{X}_t = \left[a_1(t+\Delta)\left(\varepsilon_t \hat{X}_{1t} + h_t\right) + a_0(t+\Delta)\right]dt + \varepsilon_t \beta_{1t}\left[dY_t - \left(b\hat{X}_{1t} + b_0\right)dt\right]. \tag{67}$$

$$\dot{\varepsilon}_t = a_1(t+\Delta)\varepsilon_t - \varepsilon_t a_1. \tag{68}$$

$$\dot{h}_t = a_1(t+\Delta) - \varepsilon_t a_0 + a_1(t+\Delta)h_t. \tag{69}$$

$$\dot{R}_t = a_1(t+\Delta)R_t + R_t a_1(t+\Delta)^T - \beta_t\left(b_1 v_2 b_1^T\right)\beta_t^T + \psi_1(t+\Delta)v_1(t+\Delta)b_1(t+\Delta)^T. \tag{70}$$

In case of the linear StS with the parametric Gaussian noises:

$$\begin{aligned} dX_t &= (a_1 X_t + a_0)dt + \left(c_{10} + \sum_{r=1}^{n_x} c_{1,n_y+r} X_r\right)dW_1, \\ dY_t &= (bY_t + b_1 X_t + b_0)dt + \left(c_{20} + \sum_{r=1}^{n_y} c_{2r} Y_r + \sum_{r=1}^{n_x} c_{2r,n_y+r} X_r\right)dW_2. \end{aligned} \tag{71}$$

COFc is defined by exact Eqs (Proposal 3.1.3):

$$d\hat{X}_t = \left[a_1(t+\Delta)\left(\varepsilon_t \hat{X}_1 + h_t\right) + a_0(t+\Delta)\right]dt + \varepsilon_t \beta_{1t}\left[dY_t - \left(b\hat{X}_1 + b_0\right)dt\right]. \tag{72}$$

$$\dot{\varepsilon}_t = a_1(t+\Delta)\varepsilon_t - \varepsilon_t a_1, \quad \dot{h}_t = a_0(t+\Delta) - \varepsilon_t a_0 + a_1(t+\Delta)h_t, \tag{73}$$

$$\begin{aligned} \dot{R}_t &= a_1(t+\Delta)R_t + R_t a_1(t+\Delta)^T \\ &\quad -\beta_t\left[\left(c_{20} + \sum_{r=1}^{n_y+n_x} c_{2r} m_r\right)v_1\left(c_{20}^T + \sum_{r=1}^{n_y+n_x} c_{2r}^T m_r\right)\right. \\ &\quad \left. + \sum_{r=1}^{n_y+n_x} c_{2r} v_1 c_{2s}^T k_{rs}\right]\beta_t^T + \left[c_{10}(t+\Delta) + \sum_{r=n_y+1}^{n_y+n_x} c_{1r}(t+\Delta)m_r(t+\Delta)\right]v_2(t+\Delta) \\ &\quad \times\left[c_{10}(t+\Delta)^T + \sum_{r=n_y+1}^{n_y+1} c_{1r}(t+\Delta)^T m_r(t+\Delta)\right] \\ &\quad + \sum_{r=n_y+1}^{n_y+n_x} c_{1r}(t+\Delta)v_2(t+\Delta)c_{1s}(t+\Delta)^T k_{rs}. \end{aligned} \tag{74}$$

For nonlinear StS in case of the normal StP $X_t, Y_t, \hat{X}_t$ Eqs of normal COFc (NCOFc) are defined by Proposal 3.1.1 for joint normal distribution.

3.2 Discrete and hybrid StS

Let us consider the following nonGaussian nonlinear regression StS

$$X_{k+1} = \omega_k(X_k, V_k), \quad Y_k = \omega_{1k}(X_k, Y_k, V_k) \quad (k = 1, 2 \dots). \tag{75}$$

In this case Eqs of the discrete COFc are as follows:

$$X_{k+r+1} = \delta_k \zeta_k \left(X_k, \hat{X}_k\right) + \gamma_k, \tag{76}$$

$$\delta_k = D_k B_k^{-1}, \quad \gamma_k = m_{k+r+1} - \delta_k \rho_k, \tag{77}$$

$$m_{k+r+1} = \mathrm{E}\omega_{k+r}(X_{k+r}, V_{k+r}), \tag{78}$$

$$\rho_k = \mathrm{E}\zeta_k\left(X_k, \hat{X}_k\right), \quad B_k = \mathrm{E}\left[\zeta_k\left(X_k, \hat{X}_k\right) - \rho_k\right]\zeta_k\left(X_k, \hat{X}_k\right)^T,$$

$$D_k = \mathrm{E}[\omega_{k+r}(X_{k+r}, V_{k+r}) - m_{k+r+1}]\zeta_k\left(\omega_k\left(X_k, \hat{X}_k\right)\right)^T, \tag{79}$$

$$g_{2,k,k+r}(\lambda_1, \lambda_2, \mu) = \mathrm{E}\left\{i\lambda_1^T X_k + i\lambda_2^T X_{k+r} + i\mu^T \hat{X}_k\right\}, \tag{80}$$

$$g_{2,k,k+r+1}(\lambda_1, \lambda_2, \mu) = \mathrm{E}\exp\left\{i\lambda_1^T X_k + i\lambda_2^T \omega_{k+r}(X_{k+r}, V_{k+r}) + i\mu^T \hat{X}_k\right\} \tag{81}$$

at initional condition

$$g_{2,k,k}(\lambda_1, \lambda_2, \mu) = g_{1,k}(\lambda_1 + \lambda_2, \mu). \tag{82}$$

So for the nonlinear regression StS (14) we get **Proposal 3.2.1** defined by Eqs. (75)–(82).

In case of the nonlinear autoregression discrete StS (15) we have the following Eqs of **Proposal 3.2.2:**

$$X_{k+1} = a_k(X_k) + b_k(X_k)V_k, \quad Y_k = a_{1k}(X_k, Y_k) + b_{1k}(Y_k)V_k, \tag{83}$$

$$X_{k+r+1} = \alpha_k \xi_k\left(\hat{X}_k\right) + \beta_k \eta_k\left(\hat{X}_k\right) Y_k + \gamma_k, \tag{84}$$

$$\alpha_k \kappa_{11}^{(k)} + \beta_k \kappa_{21}^{(k)} = \kappa_{01}^{(k)}, \alpha_k \kappa_{12}^{(k)}\left(\hat{X}_k\right) + \beta_k \kappa_{22}^{(k)} = \kappa_{02}^{(k)}, \tag{85}$$

$$\gamma_k = \rho_0^{(k+r+1)} - \alpha_k \rho_1^{(k)}\left(\hat{X}_k\right) - \beta_k \rho_2^{(k)} = \kappa_{02}^{(k)}, \tag{86}$$

$$\rho_0^{(k+r+1)} = \mathrm{E}a_{k+r}(X_{k+r}), \tag{87}$$

$$\rho_k = \left[\rho_1^{(k)T} \rho_2^{(k)T}\right]^T, \quad \rho_1^{(k)} = \mathrm{E}\xi_k\left(\hat{X}_k\right), \quad \rho_2^{(k)} = \mathrm{E}\eta_k\left(\hat{X}_k\right) a_{1k}(X_k), \tag{88}$$

$$B_k = \begin{bmatrix} \kappa_{11}^{(k)} & \kappa_{12}^{(k)} \\ \kappa_{11}^{(k)} & \kappa_{22}^{(k)} \end{bmatrix}, \quad \det|B_k| \neq 0, \tag{89}$$

$$\kappa_{11}^{(k)} = \mathrm{E}\left[\xi_k\left(\hat{X}_k\right) - \rho_1^{(k)}\right]\xi_k\left(\hat{X}_k\right)^T,$$
$$\kappa_{12}^{(k)} = \kappa_{21}^{(k)} = \mathrm{E}\left[\xi_k\left(\hat{X}_k\right) - \rho_1^{(k)}\right] a_{1k}(X_k)^T \eta_k\left(\hat{X}_k\right)^T,$$
$$\kappa_{22}^{(k)} = \mathrm{E}\left[\eta_k\left(\hat{X}_k\right) a_{1k}(X_k) - \rho_2^{(k)}\right] a_{1k}(X_k)^T \eta_k\left(\hat{X}_k\right)^T + \mathrm{E}\eta_k\left(\hat{X}_k\right) b_{1k}(X_k) \upsilon_k b_{1k}(X_k)^T \eta_k\left(\hat{X}_k\right)^T, \tag{90}$$

$$D_k = \begin{bmatrix} \kappa_{01}^{(k)} & \kappa_{02}^{(k)} \end{bmatrix}, \tag{91}$$

$$\begin{aligned} \kappa_{01}^{(k)} &= \mathrm{E}[a_k(X_k) - m_{k+1}]\xi_k(\hat{X}_k)^T, \\ \kappa_{02}^{(k)} &= \mathrm{E}[a_k(X_k) - m_{k+1}]a_{1k}(X_k)^T \eta_k(\hat{X}_k)^T + \mathrm{E}b_{1k}(X_k)v_k b_{1k}(X_k)^T \eta_k(\hat{X}_k)^T, \end{aligned} \tag{92}$$

$$m_{k+1} = \rho_0^{(k)}, \; \rho_0^{(k)} = \mathrm{E}a_k(X_k), \; \mathrm{E}V_k = 0, \; \mathrm{E}V_k V_k^T = v_k. \tag{93}$$

Analogously we get from Proposal 3.2.2 COFc for discrete linear StS and linear with the Gaussian parametric noises. For hybrid StS we recommend mixed algorithm based on joint normal distribution and Proposal 3.1.1.

3.3 Generalizations

Mean square results (Subsection 2.1 and 2.2) may be extended to StS described by linear, linear with the Gaussian parametric noises and nonlinear Eqs or reducible to them by approximate suboptimal and conditionally optimal methods.

Differential StS with the autocorrelated noises in observations may be also reduced to differential StS.

Special COFc algorithms based on complex statistical criteria and Bayesian creteria are developed in [11].

4. Probability modeling in SOTES

Following [3, 4] let as consider general approach for the SOTES modeling as macroscopic (multi-level) systems including set of subsystems being also macroscopic. In our case these sets of subsystems will be clusters covering that part of MP connected with aftersales production service. More precisely the set of subsystems of lower level where input information about concrete products, personal categories etc. is formed.

For typical continuous-discrete StP in the SOTES production cluster we have the following vector stochastic equation:

$$dX_t = [\varphi(X_t, t) + S(v)\rho(X_t, t)]dt + S(v)dP^0(t). \tag{94}$$

Here $P^0(t)$ being the centered Poisson StP; $\rho(X_t, t)$ being $(n_p \times 1)$ intensity of vector of StP $P(t)$, $\rho(X_t, t) = [\rho_{12}(X_t, t)\rho_{13}(X_t, t) \dots \rho_{uk}(X_t, t)]^T$; $\rho_{uk}(X_t, t)$ being intensities of streams changes of states; $\varphi(X_t, t)$ being continuous $(n_p \times 1)$ vector function of quality indicators in OPB; $S(v)$ being $(n_p \times n_\rho)$ matrix Poisson stream of resources (production) with volumes v according to the SOTES state graph. Analogously we get corresponding equations for SOTES-O and SOTES-N:

$$dY_t = [q(X_t, t) + \varphi_1(Y_t, t) + D(r)\gamma(Y_t, t)]dt + D(r)dP_1^0(t), \tag{95}$$

$$d\zeta_t = [\varphi_2(\zeta_t, t) + C(\vartheta)\mu(\zeta_t, t)]dt + C(\vartheta)dP_2^0(t), \tag{96}$$

where φ_1 and φ_2 being vector functions quality indicators in OPB for the SOTES-O and the SOTES-N; $D(r)$ being structional matrix of resources streams in the SOTES-N matrix; $\gamma(Y_t, t)$ and $D(r)$ being the intensity function and vector of $P_1^0(t)$ jumps in the SOTES-O.

In linear case when $\rho_{uk}(X_t, t) = A_\rho X_t$ Eqs. (94)–(96) for the SOTES, SOTES-O and the SOTES-N may be presented as

$$dX_t = \bar{a}(t, v)X_t dt + S(v)dP^0(t), \tag{97}$$

$$dY_t = \bar{b}(r, t)Y_t dt + \lambda(Y_t, t)X_t + D(r)dP^0(t) + \psi_1(t)d\zeta_t, \tag{98}$$

$$d\zeta_t = \bar{c}_2(\vartheta, t)\zeta_t dt + C(\vartheta)dP_2^0(t). \tag{99}$$

Here notations

$$\bar{b}_1(r, t) = b_1(t) + A_\gamma(r, t), \quad \bar{c}_2(\vartheta, t) = c_2 + A_\mu(\vartheta, t). \tag{100}$$

$A_\gamma(r, t)$, $A_\mu(\vartheta, t)$ are derived from Eqs:

$$D(r)\gamma(Y_t, t) \equiv A_\gamma(r, t)Y_t, \quad C(\vartheta)\mu(\zeta_t, t) \equiv A_\mu(\vartheta, t)\zeta_t. \tag{101}$$

At practice a priori information about SOTES-N is poor than for the SOTES and SOTES-O. So introducing the Wiener StP $W(t)$, $W_1(t)$, $W_2(t)$ we get the following Eqs:

$$dX_t = (\bar{a}X_t + a_1 Y_t + a_0)dt + S(v)dP^0(t) + \psi'(t)dW(t), \tag{102}$$

$$dY_t = \left(qX_t + b_1 Y_t + b_2\zeta_t + b_0\right)dt + D(r)dP_1^0(t) + \psi_1(t)d\zeta_t + \psi_1'(t)dW_1(t), \tag{103}$$

$$d\zeta_t = (\bar{c}_2\zeta_t + c_0)dt + C(\vartheta)dP_2^0(t) + \psi_2'(t)dW_2(t). \tag{104}$$

R e m a r k 4.1. Such noises from OTES-N may act at more lower levels OTES-O included into internal SOTES being with minimal from information point of view maximal. For highest OTES levels intermediate aggregative functions may be performed. So observation and estimation systems must be through (multi-level and cascade) and provide external noise protection for all OTES levels.

R e m a r k 4.2. As a rule at administrative SOTES levels processes of information aggregative and decision making are performed.

Finally at additional conditions:

1. information streams about OPB state in the OTES-O are given by formulae

$$Y_t = G_t(T_{st}) + T_{st} \tag{105}$$

and every StP $G_t(T_{st})$ is supported by corresponding resource (e.g. financial);

2. for SOTES measurement only external noise from SOTES-N and own noise due to error of personal and equipment are essential.

We get the following basic ordinary differential Eqs:

$$\dot{X}_t = \bar{a}X_t + a_1 G_t + a_0 + \chi_x V_\Omega \equiv L_X, 2 \tag{106}$$

$$\dot{G}_t = q(T_{st})X_t + b_2\zeta_t + \chi_g V_\Omega \equiv L_G, \tag{107}$$

$$\dot{\zeta}_t = c_2\zeta_t + c_0 + \chi_\zeta V_\Omega \equiv L_\zeta, \tag{108}$$

$$\dot{T}_{st} = bX_t + \overline{b}_1 T_{st} + b_0 + \chi_{\tau s} V_\Omega \equiv L_T. \tag{109}$$

Here $V_\Omega(t) = \left[V_x^T(t) V_g^T(t) V_\zeta^T(t) V_{st}^T(t)\right]^T$ being vector white noise, $\dim V_\Omega(t) = \left(\left(n_x + n_g + n_\zeta + n_{ts}\right) \times 1\right)$, $MV_\Omega(t) = 0$, with diagonal block intensity matrix $v_\Omega = \mathrm{diag}\left\{[v_x]\left[v_g\right][v_\zeta][v_{ts}]\right\}$, $\dim v_x(t) = \left(n_x \times n_y\right)$, $\dim v_g(t) = \left(n_g \times n_g\right)$, $\dim v_\zeta(t) = (n_\zeta \times n_\zeta)$, $\dim v_{ts}(t) = (n_{ts} \times nn_{st})$, $\chi_x, \chi_g, \chi_\zeta, \chi_{st}$ being known matrices:

$$V_x = S(v)V_P(t) + \psi'(t)V_W, \quad V_g = \psi_1(t)V_\zeta + \psi_1'(t)V_{W1}, \quad V_\zeta = C(\vartheta)V_{p2} + \psi_2'(t)V_{W2}; \tag{110}$$

$$V_P = \dot{P}_0(t), \quad V_{P1} = \dot{P}_1^0(t), \quad V_{P2} = \dot{P}_2^0(t), \quad V_{st} = V_{P1}, \quad V_W = \dot{W}(t), \quad V_{W1} = \dot{W}_1(t), \quad V_{W2} = \dot{W}_2(t). \tag{111}$$

R e m a r k 4.3. Noises V_P, V_{P1}, V_{P2} (random time moments of resources or production) in are non-Gaussian noises induced by Poisson noises in the OTES, OTES-O, OTES-N, whereas noises V_W, V_{W1}, V_{W2} (personal errors, internal noises) are Gaussian StP.

From Eqs. (110) and (111) we have the following equivalent expressions for intensities of vector $V_\Omega(t)$:

$$\begin{aligned} v_x &= S(v)\overline{\rho}S^T(v) + \psi' v_W \psi'^T, \quad v_g = \psi_1 v_\zeta \psi^T + \psi_1' v_{W1} \psi_1'^T, \\ v_\zeta &= C(\vartheta)\overline{\mu}C^T(\vartheta) + \psi_2' v_{W2} \psi_2'^T, \quad v_{ts} = D(r)\overline{\gamma}D^T(r). \end{aligned} \tag{112}$$

Here the following notations are used: $S(v)\overline{\rho}S^T(v)$, $C(\vartheta)\overline{\mu}C^T(\vartheta)$, $D(r)\overline{\gamma}D^T(r)$ being intensities of nonGaussian white noises $S(v)V_P(t)$, $D(r)V_{P1}(t)$, $C(\vartheta)V_{P2}(t)$: $\overline{\rho} = \mathrm{E}[\mathrm{diag}\rho(X_t, t)]$, $\overline{\gamma} = \mathrm{E}[\mathrm{diag}\gamma(Y_t, t)]$, $\overline{\mu} = \mathrm{E}[\mathrm{diag}\mu(\zeta_t, t)]$ being mathematical expectations of intensivity diagonal matrices of Poisson streams in the SOTES, SOTES-O, SOTES-N; v_W, v_{W1}, v_{W2} being intensivities of Gaussian white noises V_W, V_{W1}, V_{W2}. Note the difference between intensity of Poisson stream and intensity of white noise.

In case of Eqs. (106)–(109) with the Gaussian parametric noises we use the following Eqs:

$$\dot{X}_t = L_X + (\tilde{a}X_t + \tilde{a}_1 G_t)V_\Omega, \tag{113}$$

$$\dot{G}_t = L_G + \left(\tilde{q}X_t + \tilde{b}_2\zeta_t\right)V_\Omega, \tag{114}$$

$$\dot{\zeta}_t = L_\zeta + \tilde{c}_2\zeta_t V_\Omega, \tag{115}$$

$$\dot{T}_{st} = L_T + \left(\tilde{b}X_t + \tilde{b}_1 T_{st}\right)V_\Omega, \tag{116}$$

where bar means parametric noises coefficients.

At additive noises V_Ω presenting Eqs. (113)–(116) for $Z_t = [X_t \;\; G_t \;\; \zeta_t \;\; T_{st}]^T$ in form of MSL:

$$\dot{Z}_t = B_0\left(m_t^z, K_t^z, t\right) + B_1\left(m_t^z, K_t^z, t\right)Z_t + B'\left(m_t^z, K_t^z, t\right)V_t^\Omega, \tag{117}$$

we get following set of interconnected Eqs for m_t^z, K_t^z:

$$\dot{m}_t^z = B_0\left(m_t^z, K_t^z, t\right), \quad m^z(t_0) = m_0^z, \tag{118}$$

$$\dot{K}_t^z = B_1(m_t^z, K_t^z, t)K_t^z + K_t^z B_1(m_t^z, K_t^z, t)^T + B'(m_t^z, K_t^z, t)v_t^\Omega B'(m_t^z, K_t^z, t)^T, \quad K^z(t_0) = K_0^z. \tag{119}$$

Eq for $K^z(t_1, t_2)$ is given by (49).

5. Basic SOTES conditionally optimal filtering and forecasting algorithms

Proposal 5.1. *Let SOTES, SOTES-O, SOTES-N being linear, satisfy* Eqs. (102)–(104) *and admit linear filter of the form*:

$$\dot{\hat{X}}_t = (\bar{a}\hat{X}_t + a_1 G_t + a_0) + \beta_t[\dot{G}_t - (q_t \hat{X}_t + b_2\zeta_t)], \tag{120}$$

where coefficient q_t in (120) does not depend upon T_{st}. Then Eqs of optimal and conditionally optimal filters coincide with the Kalman-Bucy filter and may be presented in the following form:

$$\dot{\hat{X}}_t = \bar{a}\hat{X}_t + a_1 G_t + a_0 + R_t q_t^T v_g^{-1}[Z - (q_t\hat{X}_t + b_2\zeta_t)] \quad (Z_t = \dot{G}_t), \tag{121}$$

$$\dot{R}_t = \bar{a}R_t + R_t\bar{a}^T + v_x - R_t q_t^T v_g^{-1} q_t R_t. \tag{122}$$

Proposal 5.2. *At condition when measuring coefficient q_t depends upon $\lambda_t = \lambda(T_{st}, t)$ and admit statistical linearization*

$$\lambda(T_{st}, t) \approx \lambda_0(m_{st}, K_{st}, t) + \lambda_1(m_{st}, K_{st}, t)T_{st}^0,$$
$$\lambda_0(m_{st}, K_{st}, t) = M[\lambda(T_{st}, t)] \approx \lambda_0(m_{st}, K_{st}, t), \quad \lambda_1(m_{st}, K_{st}, t) \approx 0 \tag{123}$$

sub- and conditionally optimal filter Eqs are follows:

$$\dot{\hat{X}}_t = \bar{a}\hat{X}_t + a_1 G_t + a_0 + R_t q_0(m_{st}, K_{st})^T v_g^{-1}\{Z_t - [q_0(m_{st}, K_{st})\hat{X}_t + b_2\zeta_t]\}, \tag{124}$$

$$\dot{R}_t = \bar{a}R_t + R_t\bar{a} + v_x - R_t q_0(m_{st}, K_{st}) v_g^{-1} q_t^T(m_{st}, K_{st})R_t. \tag{125}$$

R e m a r k 5.1. Filtering Eqs defined by Proposals 5.1 and 5.2 give the m.s. square optimal algorithms nonbias of X_t for OTES at conditions of internal noises of measuring devices and external noise from OTES on measuring part of SOTES-O.

R e m a r k 5.2. Accuracy of estimation $\hat{X}_t$ depends upon not only upon noise ζ_t influencing on measuring signal but on rule and technical-economical quality SOTES criteria but on line state of resources T_{st} OPB for SOTES-O.

Using [9–11] let us consider more general SOTES than Eqs. (113)–(116) for system vector $\overline{X}_t = [X_t G_t \zeta_t T_{st}]^T$ and observation vector $\overline{Y}_t = [Y_1 Y_2 Y_3 Y_4]^T$ defined by Eqs:

$$\dot{\overline{X}}_t = (a\overline{Y}_t + a_1\overline{X}_t + a_0) + \left(c_{10} + \sum_{r=1}^{n_y} c_{1r}\overline{Y}_r + \sum_{r=1}^{n_x} c_{1,n_y+r}\overline{X}_r\right)V, \tag{126}$$

$$Z_t = \dot{\overline{Y}}_t = (b\overline{Y}_t + b_1\overline{X}_t + b_0) + \left(c_{20} + \sum_{r=1}^{n_y} c_{2r}\overline{Y}_r + \sum_{r=1}^{n_x} c_{2,n_y+r}\overline{X}_r\right)V_1. \tag{127}$$

Here a, a_0, a_1, b, b_0, b_1 and c_{ij} $\left(i = 1, 2;\ j = \overline{1, n_x}\right)$– vector–matrix functions t do not depend from $\overline{X}_t = [X_1 \ldots X_{n_x}]^T$ and $\overline{Y}_t = \left[Y_1 \ldots Y_{n_y}\right]^T$. Then corresponding algorithm of conditionally optimal filter (COF) is defined by [9–11]:

$$\dot{\hat{\overline{X}}}_t = \left(a\overline{Y}_t + a_1\hat{\overline{X}}_t + a_0\right) + \beta_t\left[\dot{Z}_t - \left(b\overline{Y}_t + b_1\hat{\overline{X}}_t + b_0\right)\right]. \tag{128}$$

For getting Eqs for β_t it is necessary to have Eqs for mathematical expectation m_t and covariance matrix K_t of random vector $Q_t = \left[X_1 \ldots X_{n_x} Y_1 \ldots Y_{n_y}\right]^T$ error covariance matrix R_t for $\tilde{X}_t = \hat{\overline{X}}_t - \overline{X}_t$. Using Eqs

$$\dot{m}_t = am_t + a_0, \tag{129}$$

$$\dot{K}_t = aK_t + K_t a^T + c_0 v c_0^T + \sum_{r=1}^{n_y+n_x}\left(c_0 v c_r^T + c_r v c_0^T\right)m_r + \sum_{r,s=1}^{n_y+n_x} c_r v c_s^T\left(m_r m_s + k_{rs}\right) \tag{130}$$

$$\left(a = \begin{bmatrix} b & b_1 \\ a & a_1 \end{bmatrix},\quad a_0 = \begin{bmatrix} b_0 \\ a_0 \end{bmatrix},\quad c_r = \begin{bmatrix} c_{2r} \\ c_{1r} \end{bmatrix}\quad \left(r = \overline{0, n_y + n_x}\right)\right), \tag{131}$$

we have the following Eq for the error covariance matrix

$$\begin{aligned}\dot{R}_t = {} & a_1 R_t + R_t a_1^T - \left[R_t b_1^T + \left(c_{10} + \sum_{r=1}^{n_y+n_x} c_{1r} m_r\right) v \left(c_{20} + \sum_{r=1}^{n_y+n_x} c_{2r}^T m_r\right)\right. \\ & \left. + \sum_{r,s=1}^{n_y+n_x} c_{1r} v c_{2r}^T k_{rs}\right]\kappa_{11}^{-1} \times \times \left[R_t b_1 + \left(c_{20} + \sum_{r=1}^{n_y+n_x} c_{2r} m_r\right) v \left(c_{10}^T + \sum_{r=1}^{n_y+n_x} c_{1r}^T m_r\right)\right. \\ & \left. + \sum_{r,s=1}^{n_y+n_x} c_{2r} v c_{2s}^T k_{rs}\right] + + \left(c_{10} + \sum_{r=1}^{n_y+n_x} c_{1r} m_r\right) v \left(c_{10}^T + \sum_{r=1}^{n_y+n_x} c_{1r}^T m_r\right) + \sum_{r,s=1}^{n_y+n_x} c_{1r} v c_{1s}^T k_{rs}.\end{aligned} \tag{132}$$

Here

$$\kappa_{11} = \left(c_{20} + \sum_{r=1}^{n_y+n_x} c_{2r} m_r\right) v \left(c_{20}^T + \sum_{r=1}^{n_y+n_x} c_{2r}^T m_r\right) + \sum_{r,s=1}^{n_y+n_x} c_{2r} v c_{2s}^T k_{rs}, \tag{133}$$

$m_t = [m_r]$ $\left(r = 1, \left(n_y + n_x\right)\right), K_t = [k_{rs}]$ $\left(r, s = \left(1, \left(m_{\overline{y}} + n_{\overline{x}}\right)\right)\right)$; V being the white nonGaussian noise of intensity v. Coefficient β_t in Eq. (127) is defined by formula

$$\beta_t = \left\{R_t b_1^T + \left(c_{10} + \sum_{r=1}^{n_y+n_x} c_{1r} m_r\right) v \left(c_{20}^T + \sum_{r=1}^{n_y+n_x} c_{2r}^T m_r\right) + \sum_{r,s=1}^{n_y+n_x} c_{1r} v c_{2s}^T k_{rs}\right\}\kappa_{11}^{-1}. \tag{134}$$

R e m a r k 5.3. In case when observations do not influence the state vector we have the following notations:

$$a = 0, \quad b = 0, \quad c_{1r} = 0, \quad c_{2r} = 0 \quad \left(r = \overline{1,4}\right), \quad n_x = 4, \quad n_y = 4;$$

$$a_1 = \begin{bmatrix} \overline{a} & \overline{a}_1 & 0 & 0 \\ q & 0 & b_2 & 0 \\ 0 & 0 & c_2 & 0 \\ b & 0 & 0 & \overline{b}_1 \end{bmatrix}, \quad a_0 = \begin{bmatrix} \overline{a}_0 \\ 0 \\ c_0 \\ b_0 \end{bmatrix}, \quad c_{10} = \begin{bmatrix} \chi_x \\ \chi_G \\ \chi_\zeta \\ \chi_{st} \end{bmatrix},$$

$$c_{1,5} = [\tilde{a} \ \tilde{a}_1 \ 0 \ 0], \quad c_{1,6} = \left[\tilde{q} \ 0 \ \tilde{b}_2 \ 0\right], \quad c_{1,7} = [0 \ 0 \ \tilde{c}_2 \ 0], \quad c_{1,8} = \left[\tilde{b} \ 0 \ 0 \ \tilde{b}_1\right]. \tag{135}$$

Proposal 5.3. *Let SOTES is described by* Eqs. (125) and (126). *Then COF algorithm is defined by* Eqs. (127)–(133).

Theory of conditionally optimal forecasting [9–11] in case of Eqs:

$$\dot{\overline{X}}_t = \left(a_1\overline{X}_t + a_0\right) - \left(c_{10} + \sum_{r=1}^{n_x} c_{1,n_y+r}\overline{X}_r\right)V_1, \tag{136}$$

$$Z_t = \dot{\overline{Y}}_t = \left(b\overline{Y}_t + b_1\overline{X}_t + b_0\right) - \left(c_{20} + \sum_{r=1}^{n_y} c_{2r}\overline{Y}_r + \sum_{r=1}^{n_x} c_{2,n_y+r}\overline{X}_r\right)V_2, \tag{137}$$

where Δ being forecasting time, V_1 and V_2 are independent nonGaussian white noises with matrix intensities v_1 and v_2, gives the following Eqs for COFc:

$$\dot{\hat{\overline{X}}}_t = \left[a_1(t+\Delta)\hat{\overline{X}}_t + a_0(t+\Delta)\right] + \beta_t\left[Z_t - \left(b\overline{Y}_t + b_1\varepsilon_t^{-1}\hat{\overline{X}}_t + b_0 - b_1\varepsilon_t^{-1}h_t\right)\right]. \tag{138}$$

where the following notations are used: $u(s,t)$ is fundamental solution of Eq: $du/ds = a_1(s)u$ at initial condition $u(t,t) = I, \quad \varepsilon_t = u(t+\Delta, t)$,

$$\beta_t = \varepsilon_t(K_x - K_{\hat{x}x})b_1^T\kappa_{11}^{-1}, \tag{139}$$

$$h_t = h(t) = \int_t^{t+\Delta} u(t+\Delta,\tau)a_0(\tau)d\tau, \quad h(t+\Delta,t) = \varepsilon_t, \tag{140}$$

$$m_x(t+\Delta) = \varepsilon_t m_x(t) + h_t = \varepsilon_t m_x + h_t. \tag{141}$$

R e m a r k 5.4. At practice COFc may be presented as sequel connection of COF, amplifier with gain $\varepsilon_t = u(t+\Delta,t)$ and summater $h_t = h(t)$:

$$\hat{\overline{X}}_t = \varepsilon_t\hat{\overline{X}}_1 + h_t, \tag{142}$$

where $\hat{\overline{X}}_t$ being the COF output or COF of current state $\overline{X}_t$.

Eq. (137) may be presented in other form:

$$\dot{\hat{\overline{X}}}_t = \left[a_1(t+\Delta)\left(\varepsilon_t\hat{X}_1 + h_t\right) + a_0(t+\Delta) + \varepsilon_t\beta_1\left[Z_t - \left(b\hat{X}_1 + b_0\right)\right]\right]. \tag{143}$$

Accuracy of COFc is defined by the following Eq:

$$\dot{R}_t = a_1(t+\Delta)R_t + R_t a_1(t+\Delta)^T - \beta_t\left[\left(c_{20} + \sum_{r=1}^{n_y+n_x} c_{2r}m_r\right)v_1\left(c_{20} + \sum_{r=1}^{n_y+n_x} c_{2r}^T m_r\right)\right.$$
$$\left. + \sum_{r,s=1}^{n_y+n_x} c_{2r}v_1 c_{2s}^T k_{rs}\right]\beta_t^T + + \left[c_{10}(t+\Delta) + \sum_{r=n_y+1}^{n_y+n_x} c_{1r}(t+\Delta)^T m_r(t+\Delta)\right]$$
$$+ \sum_{s=n^Y+1}^{n_y+n_x} c_{1r}(t+\Delta)v_2(t+\Delta)c_{1s}(t+\Delta)^T k_{rs}. \tag{144}$$

Proposal 5.5. *At conditions of Proposal 5.3 COFc is described by* Eqs. (137)–(140), *(*143*) or* Eqs. (141)–(143).

Let us consider Eqs. (94)–(96) at conditions of possible subdivision of measuring system and OPB in SOTES-O so that $q(X_t, t) = q(t)X_t$ and noise ζ_t is additive. In this case for SOTES, SOTES-O, SOTES-N Eqs. (94)–(96) may be presented in the following form:

$$\dot{X}_t = \varphi(X_t, t) + S(v)\rho(X_t, t) + \chi_x V_\Omega, \tag{145}$$

$$\dot{G}_t = q(T_{st})X_t + b_2\zeta_t + \chi_g V_\Omega, \tag{146}$$

$$\dot{\zeta}_t = \varphi_2(\zeta_t, t) + C(\vartheta)\mu(\zeta_t, t) + \chi_\zeta V_\Omega, \tag{147}$$

$$\dot{T}_{st} = \varphi_1(T_{st}, t) + D(r)\gamma(T_{st}, t) + \chi_{Ts} V_\Omega. \tag{148}$$

At condition of statistical linearization we make the following replacements:

$$S(v)\rho(X_t, t) \approx S(v)\rho_0(m_x, K_x, t) + A_{\rho 1}(m_x, K_x, t)X_t^0, \tag{149}$$

$$C(\vartheta)\mu(\zeta_t, t) \approx C(\vartheta)\mu_0(m_\zeta, K_\zeta, t) + A_{\gamma 1}(m_\zeta, K_\zeta), \tag{150}$$

$$\varphi_\zeta(\zeta_t, t) \approx \varphi_{\zeta 0} + \varphi_{\zeta 1}\zeta_t^0. \tag{151}$$

So we get the following statistically linearized expressions:

$$\varphi(X_t, t) + S(v)\rho(X_t, t) = \overline{\varphi}_{X0} + \overline{\varphi}_{X1}X_t, \quad \varphi_\zeta(\zeta_t, t) + C(\vartheta)\mu(\zeta_t, t) = \overline{\varphi}_{\zeta 0} + \overline{\varphi}_{\zeta 1}\zeta_t, \tag{152}$$

where.

$$\overline{\varphi}_{X0} = \varphi_{X0}(m_x, K_x, t) - \left[\varphi_{X1}(m_x, K_x, t) + A_{\rho 1}(m_x, K_x, t)\right]m_x + S(v)\rho_0(m_x, K_x, t),$$
$$\overline{\varphi}_{X1} = \varphi_{X1}(m_x, K_x, t) + A_{\rho 1}(m_x, K_x, t),$$
$$\overline{\varphi}_{\zeta 0} = \varphi_{\zeta 0}(m_\zeta, K_\zeta, t) - \left[\varphi_{\zeta 1}(m_\zeta, K_\zeta, t) + A_{\mu 1}(m_\zeta, K_\zeta, t)\right]m_\zeta + C(\vartheta)\mu_0(m_\zeta, K_\zeta, 0),$$
$$\overline{\varphi}_{\zeta 1} = \varphi_{\zeta 1}(m_\zeta, K_\zeta, t) + A_{\mu 1}(m_\zeta, K_\zeta, t). \tag{153}$$

Proposal 5.6. *For* Eqs. (144)–(148) *at statistical linearization conditions* Eqs. (152) and (153) *when q_t does not depend upon T_{st} suboptimal filtering algorithm is defined by Eqs:*

$$\dot{\hat{X}}_t = \overline{\varphi}_{X1}\hat{X}_t + \overline{\varphi}_{X0} + R_t q_t^T v_g^{-1}\left[Z_t - \left(q_t \hat{X}_t + b_2\zeta_t\right)\right], \tag{154}$$

$$\dot{R}_t = \overline{\varphi}_{X1}R_t + R_t\overline{\varphi}_{X1}^T + v_x - R_t q_t^T v_g^{-1} q_t R_t. \tag{155}$$

Proposal 5.7. *At conditions $\lambda_t = \lambda(T_{st}, t)$ Eqs for SOTES, SOTES-O, SOTES-N may be presented in the form:*

$$\dot{X}_t = \overline{\varphi}_{X1}X_t + \overline{\varphi}_{X0} + \chi_x V_\Omega, \tag{156}$$

$$\dot{G}_t = qX_t + b_2\zeta_t + \chi_g V_\Omega, \tag{157}$$

$$\dot{\zeta}_t = \overline{\varphi}_{\zeta 1}\zeta_1 + \overline{\varphi}_{\zeta 0} + \chi_\zeta V_\Omega, \tag{158}$$

$$\dot{T}_{st} = \overline{\varphi}_{st1}T_{st} + \overline{\varphi}_{st0} + \chi_{st}V_\Omega. \tag{159}$$

Suboptimal algorithm at condition

$$\lambda(T_{st}, t) \approx \lambda_0(m_{st}, K_{st}) \tag{160}$$

is as follows:

$$\dot{\hat{X}}_t = \overline{\varphi}_{X1}\hat{X}_t + \overline{\varphi}_{X0} + R_t\lambda_0^T v_g^{-1}\left\{Z_t - \left[\lambda_0^T(m_{st}, K_{st})\hat{X}_t + b_2\zeta_t\right]\right., \tag{161}$$

$$\dot{R}_t = \overline{\varphi}_{X1}R_t + R_t\overline{\varphi}_{X1}^T + v_x - R_t\lambda_0^T v_g^{-1}\lambda_0 R_t. \tag{162}$$

6. Peculiarities of new SOTES generations

As it was mentioned in Introduction in lower levels of hierarchical subsystems of SOTES arise information about nomenclature and character of final production and its components.

Analogously in personal LC subsystems final production systems being categories of personal with typical works and separate specialists with common works. In [1, 2] it is presented methodology of personal structuration according to categories, typical processes graphs providing necessary professional level and healthy. Analogous approach to structuration may be used to elements macroscopic subsystems of various SOTES levels. It gives possibility to design unified modeling and filtering methods in SOTES, SOTES-O, SOTES-N and then implement optimal processes of unique budget. So we get unique methodological potential possibilities for horizontal and vertical integrated SOTES.

In case of Eqs. (107)–(111) for LC subsystems in case aggregate of given personal categories defined by Eqs

$$\dot{X}_P = \overline{a}X_P + a_{1P}G_P + a_{0P} + \chi_P V_{\Omega P}, \tag{163}$$

$$\dot{G}_P = q_P(T_{sP})X_P + b_{2P}\zeta_P + \chi_{gP}V_{\Omega P}, \tag{164}$$

$$\dot{\zeta}_P = c_{2P}\zeta_P + c_{0P} + \chi_{\zeta P}V_{\Omega P}, \tag{165}$$

$$\dot{T}_{sP} = b_P X_P + b_{1P}T_{sP} + b_{0P} + \chi_{stP}V_{\Omega P}, \tag{166}$$

where index P denotes variables and parameters in personal LS. According to Section 4 the following filtering Eqs:

$$\dot{\hat{X}}_P = \bar{a}_P\hat{X}_P + a_{1p}G_P + a_{0P} + R_P q_P^T v_{gP}^{-1}\left[Z - \left(q_P\hat{X}_P + b_{2P}\zeta_P\right)\right], \tag{167}$$

$$\dot{R}_P = \bar{a}_P R_P + R_P\bar{a}_P^T + v_P - R_P q_P^T v_{gP}^{-1} q_P R_P. \tag{168}$$

Let us consider linear synergetical connection between X and X_P:

$$X_P = \kappa_1 X + \kappa_0. \tag{169}$$

Here κ_1 and κ_0 being known $(n_P \times n_x)$ and $(n_P \times 1)$ synergetical matrices. Putting (151) into Eqs. (144)–(150) we get Eqs for personal subsystem and its observation expressed by X:

$$\dot{X}_P = \bar{a}_P(\kappa_1 X + \kappa_0) + a_{1P}G_P + a_{0P} + \chi_P V_{\Omega P}, \tag{170}$$

$$\dot{G}_P = q_P(T_{sP})(\kappa_1 X + \kappa_0) + b_{2P}\zeta_P + \chi_{gP} V_{\Omega P}. \tag{171}$$

Corresponding Eqs with combined right hand for SOTES vector X are described by:

$$\dot{X} = \begin{cases} \bar{a}X + a_1 G_t + a_0 V_\Omega & \left(X_K = \overline{1, n_x}\right), \\ \bar{a}_P(\kappa_1 X + \kappa_0) + a_{1P}G_P(X) + a_{0P} + \chi_P V_{\Omega P} & \left(X_K = \overline{n_{x+1}, (n_x + n_P)}\right). \end{cases} \tag{172}$$

Analogously using Proposal 5.1 we get the Kalman-Bucy filter Eqs:

$$\dot{\hat{X}} = \begin{cases} \bar{a}\hat{X} + a_1 G_t + a_0 + R\lambda^T v_g^{-1}[Z - (\lambda\hat{X} + b_2\zeta) & (x_K = \overline{1, n_x}), \\ \bar{a}_P(\kappa_1\hat{X} + \kappa_0) + a_{1P}G_P(X) + a_{0P} + R_P\lambda_P^T v_{gP}^{-1}\{Z_P(\kappa_1\hat{X} + \kappa_0) + b_{2P}\zeta_P]\} & (x_K = \overline{n_{x+1}, n_x + n_P}), \end{cases} \tag{173}$$

$$\dot{R} = \bar{a}R + R\bar{a}^T + v_x - R\lambda^T v_g^{-1}\lambda R, \tag{174}$$

$$\dot{R}_P = \bar{a}_P R_P + R_P\bar{a}_P^T + v_P - R_P\lambda_P^T v_{gP}^{-1}\lambda_P R_P. \tag{175}$$

Eqs. (154)–(157) define Proposal 5.5 for SOTES filter including subsystems accompanying LC production and personal taking part in production LC.

Remark 6.1. Analogously we get Eqs for SOTES filter including for financial subsystem support and other subsystems.

Remark 6.2. Eqs. (173) and (174) are not connected and may be solved a priori.

7. Example

Let us consider the simple example illustrating modeling the influence of the SOTES-N noise on rules and functional indexes of subsystems accompanying LC production, its filtration and forecasting. System includes stocks of spare parts (SP), exploitation organization with park of MP and together with repair organization (**Figure 1**).

At initial time moment necessary supplement provides the required level of effective exploitation at time period $[0, T]$. Let consider processes in ASS connected with one type of composite parts (CP) in number N_T. During park exploitation CP failures

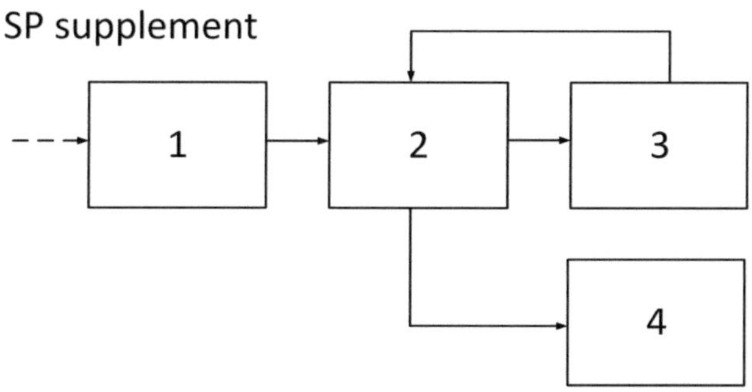

Figure 1.
After Sale system (ASS) for MP.

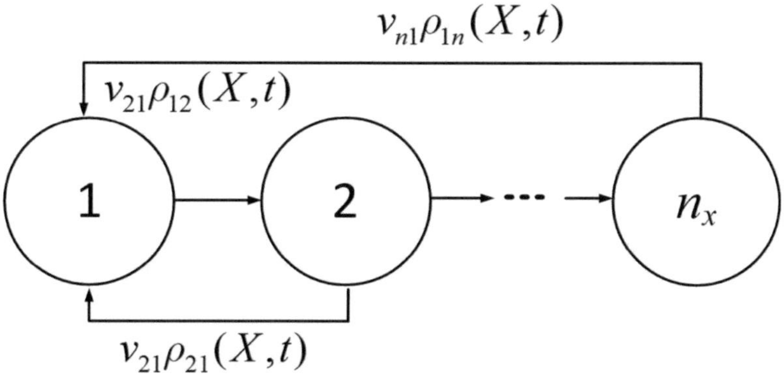

Figure 2.
Graph of production state.

appears. Non-repaired CP are or repaired again and returned into exploitation or whiting off. If the level of park readiness in exploitation is less the critical the repaired MP are taken from stocks.

In graph (**Figure 1**) the following notations are used: (1) being in stocks in number X_1, (2) exploitation in number X_2, (3) repair in number X_3, (4) witting off in number X_4. Using **Figure 2** we $n_x = 4$; transitions: $1 \to 2v = 1$ (for the Poisson stream $\rho_{12}X_1$); $2 \to 3 - \rho_{23}X_2$; $2 \to 4 - \rho_{24}X_2$; $3 \to 2$; number of transitions is equal to $n_p = 4$. As index of efficiency we use the following coefficient of technical readiness [1, 2]:

$$K_{tr}(T) = \frac{1}{T}\int_0^T \frac{X_2(\tau)d\tau}{N_T} = \frac{1}{TN_T}\int_0^T X_2(\tau)d\tau, \tag{176}$$

where

$$N_T = X_2 + X_3 + X_4 \tag{177}$$

Being constant number of CP of park in exploitation.

Note that the influence of the SOTES-N on SOTES and SOTES-O is expressed in the following way: system noise ζ_t as factor of report documentation distortion leads to fictive underestimated $\overline{K}_{TR}(t)$. In case when relation

$$\overline{K}_{TR}(t) \geq \overline{K}^*_{TR}(t) \tag{178}$$

is breaked down ($\overline{K}^*_{TR}(t)$ being critical of floating park level) stocks will give necessary CP amount. So we receive the possibility to exclude defined CP amount from turn over.

Finally let us consider filtering and forecasting algorithms of ASS processes for exposition of noise ζ_t.

Solution.

1. For getting Eq for $\overline{K}_{TR}(t)$ we assume $X^*_{TR}(t) = \overline{K}_{TR}(t)$ and take into account Eqs. (106)–(109). So have the following scalar Eqs:

$$\dot{X}_{TR}(t) = \frac{1}{TN_T} X_2(t),$$

$$\dot{X}_1 = -\rho_{12}X_1, \quad \dot{X}_2 = \rho_{12}X_1 - (\rho_{23} + \rho_{24})X_2, \quad \dot{X}_3 = \rho_{23}X_2 - \rho_{32}X_3, \quad \dot{X}_4 = \rho_{24}X_2. \tag{179}$$

In our case $a_1 = a_0 = 0, \quad V_X = 0$ and Eq. (10.4) may be presented in vector form

$$\dot{X} = aX \tag{180}$$

where

$$a = \begin{bmatrix} 0 & 0 & (TN_T)^{-1} & 0 & 0 \\ 0 & -\rho_{12} & 0 & 0 & 0 \\ 0 & \rho_{12} & -(\rho_{23} + \rho_{24}) & 0 & 0 \\ 0 & 0 & \rho_{23} & -\rho_{32} & 0 \\ 0 & 0 & \rho_{24} & 0 & 0 \end{bmatrix}. \tag{181}$$

At practice the reported documentation is the complete of documents containing SP demands from stock and acknowledgement SP documents. So noise ζ_t acts only if its realization take part delivery and acquisition sides. This is the reason to name this noise as system noise and carried out by group of persons.

2. For setting Eqs for electronic control system we use of the following type Eq. (108)

$$\dot{G} = qX + \zeta + \chi_g V_\Omega \tag{182}$$

where $V_\Omega = [V_{TR}V_1V_2V_3V_4]^T$ being noises with intensity v_g; $\lambda = [\lambda_{TR}\lambda_1\lambda_2\lambda_3\lambda_4]^T$ being ecoefficiency of measuring block. In scalar form Eq. (181) may be presented as

$$\begin{aligned} &\dot{G}_{TR} = \lambda_{TR}X_{TR} + \zeta + V_{TR}, \quad \dot{G}_1 = \lambda_1 X_1 + V_1, \quad \dot{G}_2 = \lambda_2 X_2 + V_2, \\ &\dot{G}_3 = \lambda_3 X_3 + V_3, \quad \dot{G}_4 = \lambda_4 X_4 + V_4. \end{aligned} \tag{183}$$

3. Algorithm for noise ζ_t description depends on participants. In simple case we use error

$$\delta X_{TR}(t) = X_{TR}(t) - \overline{K}^*_{TR}(t). \tag{184}$$

In this case we get lag in G_{TR} measurement on variable ζ_t:

$$\zeta_t = b_2(X_{TR})|X_{TR}(t) - \overline{K}^*_{TR}|. \tag{185}$$

By the choice of coefficient b_2 necessary time temp of documentation manipulation may be realized.

4. Using Eqs of Proposals 5.1 and 5.2 we get the following matrix filtering Eqs for system noise ζ_t on background of measuring noise V_{TR}

$$\dot{\hat{X}} = a\hat{X} + R\lambda^T v_g^{-1}\left[Z - \left(\lambda\hat{X} + \zeta\right)\right], \tag{186}$$

$$\dot{R} = aR + Ra^T - R\lambda^T v_g^{-1}\lambda R. \tag{187}$$

at $Z = \dot{G}$, $\zeta = [\zeta_t \;\; 000]^T$.

R e m a r k 7.1. Realization of the described filtering solutions for internal noises needs a priori information about basic OTES characteristics. So we need special methods and algorithms.

5. Finally linear COFc is defined by Eqs. (137)–(104) for various forecasting times Δ.

R e m a r k 7.2. In case of SOTES with two subsystems using Eqs. (172)–(174) we have the following Kalman-Bucy filter:

$$\dot{\hat{X}} = \begin{cases} a\hat{X} + R\lambda^T v_g^{-1}\left[Z - \left(\lambda\hat{X} + \zeta\right)\right], & X_K = \overline{1, n_x}; \\ a_P\left(\kappa_1\hat{X} + \kappa_0\right) + R_P\lambda_P^T v_{gP}^{-1}\left\{Z_P - \left[\lambda_P\left(\kappa_1\hat{X} + \kappa_0\right)\zeta_P\right]\right\}, & X_K = \overline{n_{xt}, n_x + n_P}, \end{cases}$$

where ζ_P being noise acting on the functional index of personal attendant subsystem.

These results are included into experimental software tools for modeling and forecasting of cost and readiness for parks of aircraft [1, 2].

8. Conclusion

For new generations of synergetical OTES (SOTES) methodological support for approximate solution of probabilistic modeling and mean square and forecasting filtering problems is generalized. Generalization is based on sub- and conditionally optimal filtering. Special attention is paid to linear systems and linear systems with the parametric white Gaussian noises.

Problems of optimal, sub- and conditionally optimal filtering and forecasting in product and staff subsystems at the background noise in SOTES are considered. Nowadays for highly available systems the problems of creation of basic systems engineering principles, approaches and information technologies (IT) for SOTES from modern spontaneous markets at the background inertially going world economics crisis, weakening global market relations at conditions of competition and counteraction reinforcement is very important. Big enterprises need IT due to essential local and systematic economic loss. It is necessary to form general approaches for stochastic processes (StP) and parameters estimation (filtering, identification, calibration etc) in

SOTES at the background noises. Special observation SOTES (SOTES-O) with own organization-product resources and internal noise as information from special SOTES being enact noise (SOTES-N). Conception for SOTES structure for systems of technical, staff and financial support is developed. Linear, linear with parametric noises and nonlinear stochastic (discrete and hybrid) equations describing organization-production block (OPB) for three types of SOTES with their planning-economical estimating divisions are worked out. SOTES-O is described by two interconnected subsystems: state SOTES sensor and OPB supporting sensor with necessary resources. After short survey of modern modeling, sub- and conditionally optimal filtering and forecasting basic algorithms and IT for typical SOTES are given.

Influence of OTES-N noise on rules and functional indexes of subsystems accompanying life cycle production, its filtration and forecasting is considered.

Experimental software tools for modeling and forecasting of cost and technical readiness for parks of aircraft is developed.

Now we are developing presented results on the basis of cognitive approaches [12].

Acknowledgements

The authors would like to thank Russian Academy of Sciences and for supporting the work presented in this chapter.

Authors much obliged to Mrs. Irina Sinitsyna and Mrs. Helen Fedotova for translation and manuscript preparation.

Author details

Igor N. Sinitsyn* and Anatoly S. Shalamov
Federal Research Center "Computer Sciences and Control of Russian Academy of Sciences", Moscow, Russia

*Address all correspondence to: sinitsin@dol.ru

References

[1] Sinitsyn IN, Shalamov AS. Lectures on Theory of Integrated Logistic Support Systems. 2nd ed. Moscow: Torus Press; 2019. p. 1072 (in Russian)

[2] Sinitsyn IN, Shalamov AS. Probabilistic modeling, estimation and control for CALS organization-technical-economic systems – Chapter 5/in book "Probability, Combinatorics and Control". In: Kostogryzov A, Korolev V, editors. London, UK: IntechOpen; 2020. p. 117-141. DOI: 10.5772/intechopen.79802

[3] Sinitsyn IN, Shalamov AS. Optimal estimation and control in stochastic synergetic organization-technical-economic systems. Filtering in product and staff subsystems at background noise (I). Highly Available Systems. 2019;**15**(4):27-48. DOI: 10.18127/j20729472-201904-04 (in Russian)

[4] Sinitsyn IN, Shalamov AS. Optimal estimation and control in stochastic synergetic organization-technical-economic systems. Filtering in product and staff subsystems at background noise (II). Highly Available Systems. 2021;**17**(1):51-72. DOI: 10.18127/j20729472-202101-05 (in Russian)

[5] Sinitsyn IN, Shalamov AS. Problems of estimation and control in synergetic organization-technical-economic systems. In: VII International Conference "Actual Problems of System and Software". Moscow; 2021 (in print)

[6] Haken H. Synergetics and Introduction. Springer Ser. Synergistics. Vol. 3. Berlin, Heidelberg: Springer; 1983

[7] Haken H. Advanced Synergetics. Springer Ser. Synergistics. Vol. 20. Berlin, Heidelberg: Springer; 1987

[8] Kolesnikov AA. Synergetical Control Theory. Taganrog: TRTU: M.: Enegroatomizdat; 1994. p. 538 (in Russian)

[9] Pugachev VS, Sinitsyn IN. Stochastic Systems. Theory and Application. Singapore: World Scientific; 2001. p. 908

[10] Sinitsyn IN, editor. Academic Pugachev Vladimir Semenovich: To 100 Annyversary. Moscow: Torus Press; 2011. p. 376 (in Russian)

[11] Kalman SIN. Pugachev Filters. 2nd ed. Vol. 2007. Moscow: Logos; 2005. p. 772 (in Russian)

[12] Kostogryzov A, Korolev V. Probabilistic Methods for cognitive solving of some problems in Artifial intelligence system – Chapter 1/in book "Probability, Combinatorics and Control". In: Kostogryzov A, Korolev V, editors. London, UK: IntechOpen; 2020. p. 3–34. DOI: 10.5772/intechopen.89168.

Section 4

Probability and Statistical Methods

Chapter 4

Probabilistic Predictive Modelling for Complex System Risk Assessments

Andrey Kostogryzov, Nikolay Makhutov, Andrey Nistratov and Georgy Reznikov

Abstract

The risks assessment is described by the action of estimating the probability distribution functions of possible successes or failures of a system during a given prediction period. Typical probabilistic predictive models and methods for solving risks prediction problems are described, and their classification is given. Priority development directions for risks prediction in standard system processes and their implementation procedures are proposed. The reported examples demonstrate the effects and interpretation of the predictive results obtained. Notes: 1. System is a combination of interacting elements organized to achieve one or more stated purposes (according to ISO/IEC/IEEE 15288 "Systems and software engineering—System life cycle processes"). 2. Risk is defined as the effect of uncertainty on objectives considering consequences. An effect is a deviation from the expected — positive and/or negative (according to ISO Guide 73).

Keywords: prediction, method, model, probability, risk

1. Introduction

Systems are subject to various risks throughout their life cycles despite their successful design and effective operation. That is why mathematics and system performance prediction have been closely interrelated since the ancient times. There is no doubt in the design and the maintenance of the world-famous wonders, astonish modern man. The preservation of these wonders was entirely based on predictive methods using existing mathematical approaches by that time. With the advent of probability theory, this relationship has become even closer. Currently, various classical mathematics and probabilistic methods are often used to solve complex engineering problems.

If for the layman probability is still associated with divination on daisies, then for specialists these methods have long become powerful tools in predicting success or failure, proactive management, and achieving the desired effects. Risk predictive assessments are practiced in various industrial sectors, for example, fuel and energy, pharmaceutical, mining, metallurgical, chemical, communication and information, dispatch centers, etc. [1–32]. Hundreds of universities and other scientific organizations are involved in probabilistic research activities connected to risk prediction. By now it is possible to clearly trace the activities chain in a predictive approach: "From uncertainties formalization – to probabilistic modelling", "From probabilistic modelling – to reasonable control", "From reasonable control – to achievable effects" and "From achievable effects –to sustainable harmony". It means that predictive probabilistic concepts meet the main analytical challenges in the eternal aspirations to go from uncertainties formalization" to "sustainable harmony", see **Figure 1**.

Thousands of mathematicians are currently involved in risk prediction R&D activities. It is unfortunately impossible to mention all the running developments. This chapter will focus on:

- some generalizations and thoughts regarding the variety of the existing risk prediction probabilistic approaches;
- the formulation of the goals and objectives of the probabilistic methods throughout the life cycle of various systems;

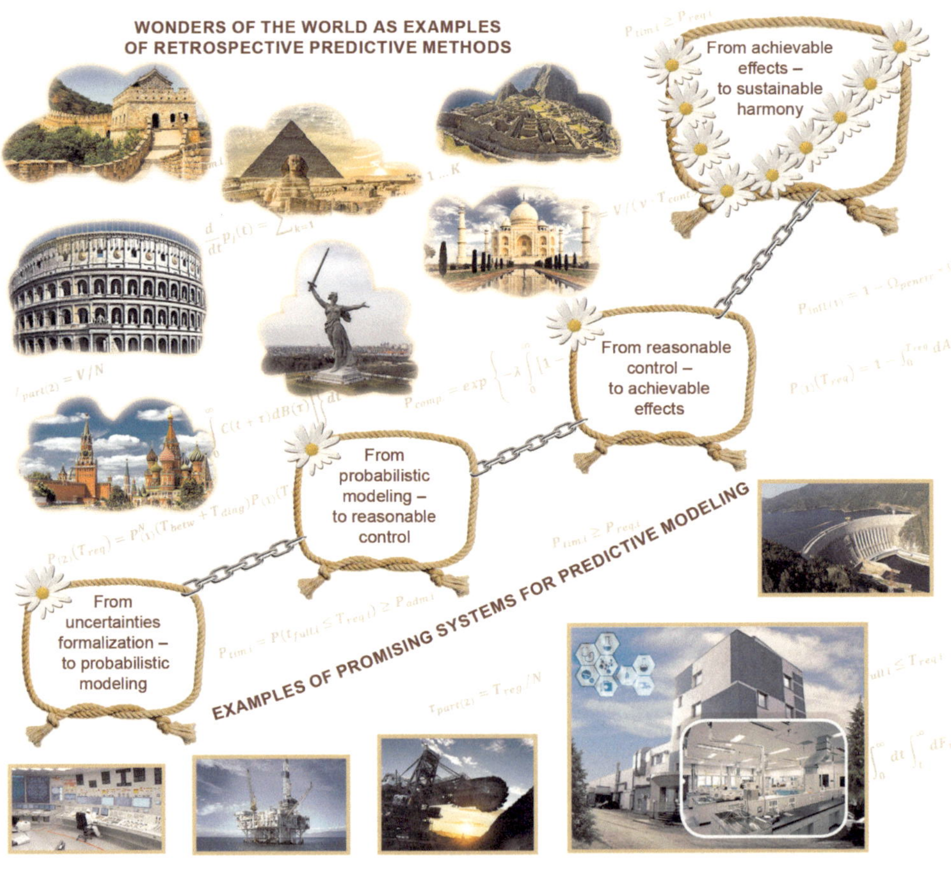

Figure 1.
The eternal aspirations: "From uncertainties formalization–to sustainable harmony."

- the description of the general risk prediction probabilistic approach;
- the essence of the probabilistic concepts considering the acceptable risk notion;
- some original probabilistic models;
- the analytical methods of risks integrating for standard processes;
- some optimization problem statements for rational proactive actions;
- some examples of practical applications (illustrating some scientific and technical possibilities for solving real engineering problems);
- the expected achievable effects.

2. Goals and objectives

In general, risk prediction is associated with the achievement of pragmatic goals and solving the analytical problems of systems rational concept (conceptual design), development, utilization, and support. Pragmatic system goals may be:

- improving the efficiency of the implementation of the state and/or corporate strategy in the economy;
- improving the safety and sustainability of the region's development, ensuring socio-economic, pharmaceutical, medical, and biological safety of the region;
- ensuring the protection of the population and territories from natural and man-made risks, etc.

In turn, the following objectives require risk predictive capabilities:

- to predict the mean residual time before the next operational abnormality;
- to ensure the effective operation and development of complex engineering, energy, transport, and communication systems;
- to ensure the security of critical infrastructure, information, and information-psychological security;
- to ensure energy and industrial safety, technical diagnostics and resource management for critical facilities and systems;
- to ensure the safety of railway, aviation and water transport;
- to develop critical technologies (for example, information and cognitive technologies; energy technologies of the future; technologies for monitoring and predicting the state of the environment and equipment; technologies for

exploration and development of mineral deposits and their extraction; and technologies for preventing and eliminating natural and technogenic hazards), etc.

A review of the numerous existing mathematical approaches allows us to make the following generalization –the main goals of applying probabilistic prediction are connected with (see **Figure 2**):

- an analysis of opportunities, achievable quality, safety and efficiency;
- a rationale for achieving desirable characteristics and acceptable conditions;
- an optimization of systems performance and processes;
- exploring new ideas and innovative concepts.

The enlarged classification of methods, using the probabilistic risk predictive models (including the proposed models), is presented in **Table 1**. These methods are used for solving various objectives during system life cycle.

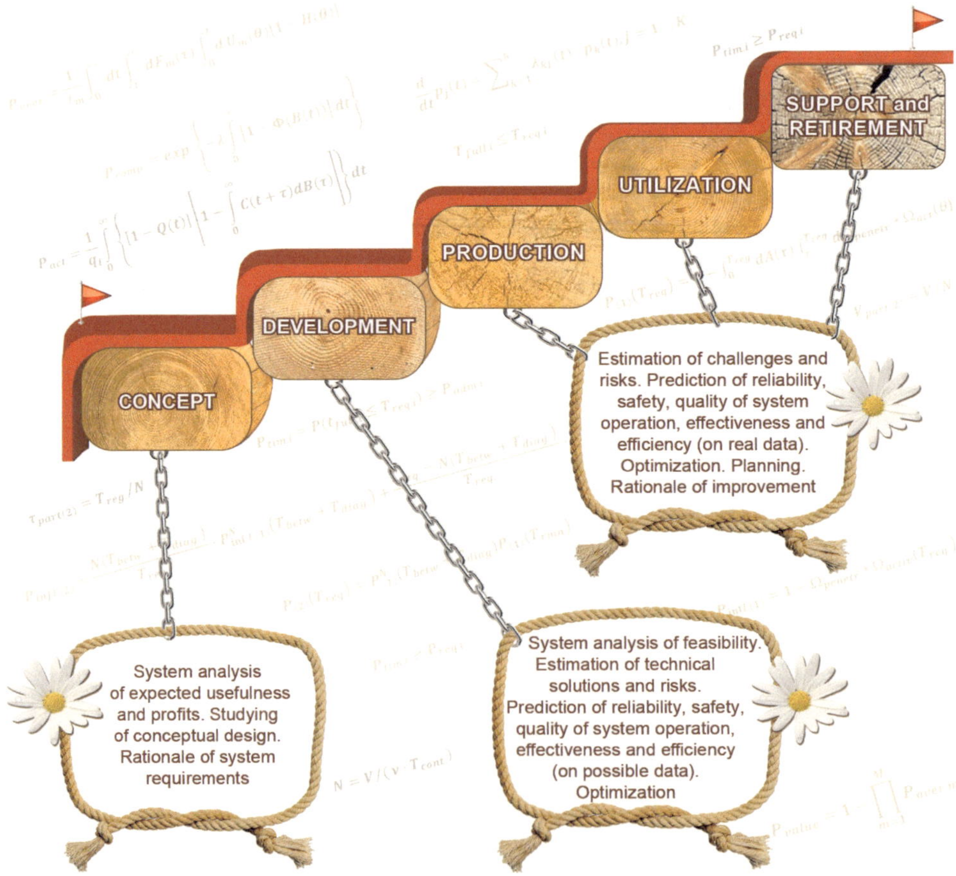

Figure 2.
Generalization of goals and objectives throughout the system's life cycle that require risk probabilistic-predictive models.

Stages in life cycle (see, for example, ISO/IEC/IEEE 15288). The problems which are due to be solved by risks prediction	Methods, connected with:			
	an analysis of opportunities, achievable quality, safety, efficiency	a rationale of achieving desirable characteristics and acceptable conditions	optimization of systems and processes	finding and researching of new ideas and concepts
Concept stage. Problems connected with a system analysis of expected usefulness and profits, studying of system creation, the rationale of system requirements and acceptable conditions	Methods for estimating critical measures. Methods for probabilistic risk prediction	Methods for estimating critical measures. Methods for probabilistic risk prediction	Methods for optimization, considering risks prediction	Methods to analyze possible effects. Methods for probabilistic risk prediction.
Development stage. The problems connected with a system analysis of feasibility, the estimations of technical solutions and risks, the prediction of reliability, safety, a quality of system operation, effectiveness and efficiency (on possible data), optimization	Measurements. Methods for estimating critical measures. Methods for probabilistic risk prediction	Measurements. Methods for estimating critical measures. Methods for probabilistic risk prediction	Methods for optimization, considering risks prediction	Methods to analyze possible effects. Methods for probabilistic risk prediction.
Production stage. The problems connected with an estimation of challenges and risks, the prediction of reliability, safety, quality of system operation, effectiveness and efficiency (on real data), optimization, planning, rationales for improvement	Measurements. Methods for estimating critical measures	Measurements. Methods for estimating critical measures	Methods for production optimization, considering risks prediction	Methods for estimating critical measures.
Utilization stage. The problems connected with an estimation of challenges and risks, the prediction of reliability, safety, a quality of system operation, effectiveness and efficiency (on real and possible data), optimization, planning, rationale of improvement	Measurements. Methods for estimating critical measures. Methods for probabilistic risk prediction	Measurements. Methods for estimating critical measures. Methods for probabilistic risk prediction	Methods for optimization, considering risks prediction	Methods to analyze possible effects. Methods for estimating critical measures. Methods for probabilistic risk prediction.
Support and retirement stages. The problems connected with an estimation of challenges and risks, the predicting of reliability, safety, quality of system operation, effectiveness and efficiency (on real and possible data), optimization, planning, rationale of improvement (in part concerning)	Measurements. Methods for estimating critical measures. Methods for probabilistic risk prediction	Measurements. Methods for estimating critical measures. Methods for probabilistic risk prediction	Methods for optimization, considering risks prediction	Methods for estimating critical measures. Methods for probabilistic risk prediction

Table 1.
The enlarged classification of methods, using risk probabilistic-predictive models.

3. Conceptual probabilistic-predictive approach

The solution of problems in the system life cycle [6–8, 9, 14] is considered by the example of a complex system, designated as (S-N-T)-system and covering: social sphere S (person, society, state and world community); natural sphere N (earth and space); techno-sphere T (man-made infrastructures and life support facilities).

In general, solving problems using a probabilistic-predictive approach includes:

- obtaining new knowledge about the fundamental laws of the operation and development of (S-N-T)-system in time and defining the probabilistic expressions and their parameters;

- formation of specific goals, concepts, and conditions in the system life cycle (with the construction of fault trees and event trees, as well as risk matrices for infrastructures and facilities), operation (including quality assurance, safety, efficiency) and development and their integration (taking into account certain priorities) for each of these areas (S, N, T) and (S-N-T)-system as a whole;

- rationalizing and building scientifically based predictions of the (S-N-T)-system development, as well as each of the constituent spheres (S, N, T), to achieve certain goals during the life cycle and to retain the critical parameters and measures within acceptable limits;

- rationalizing means, methods, and technologies for sustainable development of the (S-N-T)-system based on new knowledge and reasonable predictions;

- planning of rational (S-N-T)-system process management taking into account feedbacks;

- practical implementation and control (on-line and off-line) of the predictions and plans fulfillment for the operation and sustainable development of the (S-N-T) system, taking into account social, natural, and man-made hazard-exposure uncertainties.

When planning and implementing these actions, the following should be taken into account:

- the complexity and uncertainty of (S-N-T)-system probabilistic-predictive models, many challenges and threats leading to a deterioration of the system integrity, the effects of damaging factors, and the decrease in the survivability of the system;

- the time dependence of interrelations between spheres and components of the system, subsystems and significant elements, vulnerabilities and admissible limits for the (S-N-T)-system states in the conditions of possible challenges and threats;

- the need to categorize and classify (S-N-T)-system according to the level of importance and criticality in order to achieve goals throughout the life cycle and to retain critical parameters and measures within acceptable limits.

The random time variables τ considered in the predicted risk $R(\tau, t)$ does simultaneously take into account the probabilities $P(\tau, t)$ of the threats' occurrence and activation, and also the associated damages $U(\tau, t)$. For example, the random time variable τ may be defined as the time between successive losses of element integrity (see details in sections 4 and 5). Here the prediction period t (which in general is also subject to justification) is dependent on the basic measures, designed to characterize the uncertainties and complexity of (S-N-T)-system, and conditions for solving the analytical problems.

The source of risks regarding the (S-N-T) system has been and remains: human (human factor); nature with its own range of threats; and techno-sphere with its inherent hazards. They are the determinants of the reliability (including aging and degradation of technical means), other quality measures (including the quality of the information used), and the safety and efficiency of the system. This makes it possible to determine risk as functionals:

$$R(\tau, t) = F\{P(\tau,\ t), U(\tau,\ t)\} = F\{R_S(\tau,\ t), R_N(\tau,\ t), R_T(\tau,\ t)\}.$$

In practice, risks are estimated by the dimensionless probability of an elementary event during a period of time, comparing possible damage to it, or by the probabilistic expectation of damage (as the probabilistic multiplication of the possible damage on the probability of damage), or by the frequency of damage, etc. In turn, the magnitude of damages can be estimated in economic indicators (financial), areas of contamination, losses in case of accidents, etc.

For example, formalization of such limitations may be presented as follows:

$$R(\tau, t) \le Radm(\tau, t), Radm(\tau, t) \succ 0.$$

Then a safety $S(\tau, t)$ for (S-N-T)-system can be expressed in terms of risks: $S(\tau, t) \le Radm(\tau, t) - R(\tau, t)$. Safety is maintained if and only if $S(\tau, t) \ge 0$.

To ensure that the quality, safety and sustainable development of the (S-N-T)-system are in the acceptable risk zones. Thus, it is necessary to implement a set of actions with the economic costs expected to reduce risks to an acceptable level.

Examples of the applicability of this approach are proved in many industrial sectors such as nuclear, thermal and hydraulic power plants; the largest installations of oil and gas chemistry; the unique space station, aviation, sea and land transport; large-scale offshore energy resources development facilities [7], etc.

4. The essence of probabilistic concepts

The risk predictive approaches, used by system analysts, are based on classical probability theory. Generally, a probabilistic space (Ω, B, P) should be created per system (see for example [1–6, 9–14]), where: $\boldsymbol{\Omega}$ – is a finite space of elementary events; B – is a class of subspaces in $\boldsymbol{\Omega}$ -space with the properties of $\boldsymbol{\sigma}$-algebra; $\boldsymbol{P}$ – is a probability measure on a space of elementary events $\boldsymbol{\Omega}$. Because $\boldsymbol{\Omega} = \{\omega_k\}$ is finite, there is enough to establish a correspondence $\omega_k \to \boldsymbol{p_k} = \boldsymbol{P}(\omega| \supset| \boldsymbol{k}|)$ in which $\boldsymbol{p_k} \ge \mathbf{0}$ and $\sum_k \boldsymbol{p_k} = \mathbf{1}$. Briefly, the initial formulas in mathematical form for original models (which are used in practice) are given in **Appendices A** and **B**.

Note. Some cases of a limited space of elementary events see in Section 6. The results of modelling are related only to introduced elementary events and specific interpretation, the results of the probabilistic prediction can not describe future exact events (place, time and other detailed characteristics).

The next interconnected concepts 1–7 are proposed for probabilistic predictive modelling.

Concept 1 is concerning the probability distribution function (PDF) $P(\tau \leq t)$ (see for example [1–6, 9–14] etc.) for a continuous random variable of time τ. $P(\tau \leq t)$ is a non-decreasing function $P(t)$ whose value for a given point $t \geq 0$ can be interpreted as the probability that the value of the random variable τ is less or equal to the given time t. Regarding risk prediction, the given time t indicates the prediction period. Additionally, $P(t) = 0$ for $t \leq 0$, and $P(t) \to 1$ for $t \to \infty$. From a decision-making standpoint, the problem is to determine the probability of system "success" and/or "unsuccess" during the given prediction period T_{req} (for example, a risk of "failure" considering consequences). This probability is a value for a point t = T_{req}, and a PDF is due to be built for modelling the system's operational states with the time.

Concept 2. The processes, connected with data processing should provide the required system operational quality (because the system performs functions by logical reasoning based on data processing). The corresponding probabilistic methods should be appropriate for the assessment of the quality of the used information [6–8, 9–14, 28–31].

Concept 3. The PDF should establish the analytical dependence between the input parameters to allow solving direct and inverse problems necessary for the rational management of the system operation. For example, the PDF P(t) describing the random time τ between successive losses of integrity of a system may be an analytical exponential approximation of a simple system element, i.e. $P(t) = 1 - exp\,(-\lambda t)$, where λ is the frequency of failures (losses of element integrity per unit of time). At the same time, the frequency of failures may be considered as a sum of frequencies of different types of failures because of various specific failure reasons—for example, failure from equipment λ_1, or from natural threats λ_2, or from "human factor" λ_3 and so on. For this use case, PDF may be presented as $P(t) = 1 - exp\,[-(\lambda_1 + \lambda_2 + \lambda_3 + \ldots)t]$, if and only if all the implied failures are independent. Then if the PDF $P(t)$ is built in dependence on different parameters and if an admissible probability level for acceptable risk is given then the inverse problem may be solved analytically—see also Section 7.

Notes. 1 System integrity is defined as such system state when system purposes are achieved with the required quality. 2. The rationale for exponential approximation choice in practice see for example in [6, 9, 14, 28–31].

Concept 4. Acceptable adequacy must be ensured. It means the consideration of several essential system parameters on which "success" or "failure" of the system operation is dependent. For example, today the way for risks prediction based on only one parameter – frequency of failures λ – is common. For this case, the exponential PDF can be used—see **Figure 3**. But the required acceptable adequacy is not always proven.

For exponential approximation the frequency of failures λ is connected with the hypothesis: "No failures during the given time with a probability less than the given admissible probability P_{adm}>0". This is always the case if the failure frequency is constant with time. For this case, the given prediction time must be no more than $t_{req} = 1/\lambda_{adm}$, here $\lambda_{adm} = \frac{-ln\,(1-P_{adm})}{t_{req}}$. That may not be often an accurate engineering estimation because many systems' capabilities and operation conditions are ignored [9, 14]. In **Figure 3**, this case is explained on the timeline.

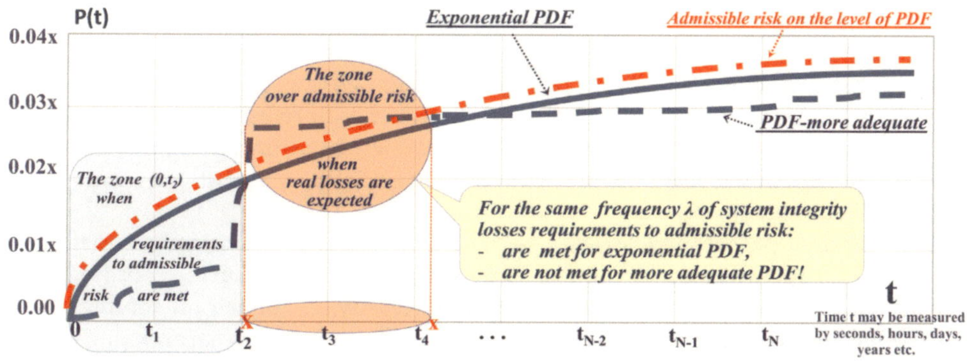

Figure 3.
Probabilistic risk, approximated by a more adequate PDF P(t), *in comparison with the existing representation of exponential PDF (both connected with the same* λ*), and admissible risk, imaginary by exponential PDF, connected with* λ_{adm}.

For different approaches and discussions, devoted to adequacy, see for example the work in [33]. In that case, the diagnostic approach to evaluate the predictive performance is based on the paradigm of maximizing the sharpness of the predictive distributions. After calibration, one obtains an assessment and ranking of the probabilistic predictions of wind speed at the Stateline wind energy centre in the US Pacific Northwest. In [34], the approach is illustrated by examples connected with "human factors". For specific systems, the topic of improving the adequacy of the prediction will always remain relevant.

Concept 5. A complex system includes subsystems and/or components (system elements), the probabilistic approach must allow a generation of probabilistic predictive models to predict the system's operational performance and its dependence on different uncertainty conditions. In general, predictive models must consider system complexity, the diagnostics of system's integrity, the monitoring of the diagnostics, the recovery from loss integrity of every system component and the quality of the used information. The adequate PDF must be the output of the probabilistic-predictive models (see also **Appendix A**).

Concept 6. The input for the probabilistic-predictive models must be based on real and other possible data (subjective data, meta-data, etc.) considering the system operational specifications and the supporting actions. These may be also hypothetical data for research purposes.

Concept 7. The specific problems of optimization must be solved considering risks prediction results (including optimization in real time). The given time for prediction should be defined so to be in real system operation time to allow taking rational proactive actions.

5. The description of some original probabilistic models

For modelling modern and future systems, taking into account their specifications, it makes sense to distinguish between the intellectual part, where uncertainties are associated with information gathering, processing and production for decision-making, and the technical part, where there is no significant dependence on the high quality of the current information.

5.1 About system operational information quality

The created models [6–8, 9–14, 28–31] help to implement concepts 1 and 2. In general, operational information quality is connected with requirements for reliable and timely producing complete, valid and/or confidential information, if needed. The gathered information is used for proper system specificity. The abstract view of such quality is illustrated in **Figure 4**.

The proposed probabilistic predictive models to assess the information quality are described in **Appendix A**. The models cover the predictive measures according to the abstract information quality metrics in **Figure 4**. It may be applied for solving problems connected with decision-making on the base of information gathering, processing and production.

5.2 About "black box" formalization to predict "failure" risks

The models below help to implement concepts 1, 3 and 4 [6, 9, 14–31]. In general, successful system operation is connected with counteractions against various system integrity loss hazards (of social, natural and technogenic origins) throughout system operation timeline. There are considered two general technologies formalized to predict "failure" risks. Both technologies are briefly described below.

Technology 1 is based on a periodic diagnostic of system integrity policy. It is carried out to detect system functional abnormalities or degradations that may result in a system loss of integrity. The system loss of integrity can be detected only as a result of diagnostics. Dangerous influence on system is logically acted step-by-step: at first, a danger source penetrates into system and then after its activation begins to influence. System integrity can not be lost before penetrated danger source is activated. A danger is considered to be realized only after danger source has influenced on system.

Notes: 1. For example, for new steel structures, time before the appearance of critical erosion from rust can be considered as the source penetration time, activation time is the time before unacceptable structural damage occurs due to this rust. 2. Regarding a degradation of technical system the input time of danger source penetration tends to zero. 3. For special research cases of cyberattacks the term "Loss of Integrity" may be logically replaced by the term "functional abnormalities".

Technology 2, additionally to technology 1, implies that system integrity is monitored between diagnostics by operator. An operator may be a man or a special artificial

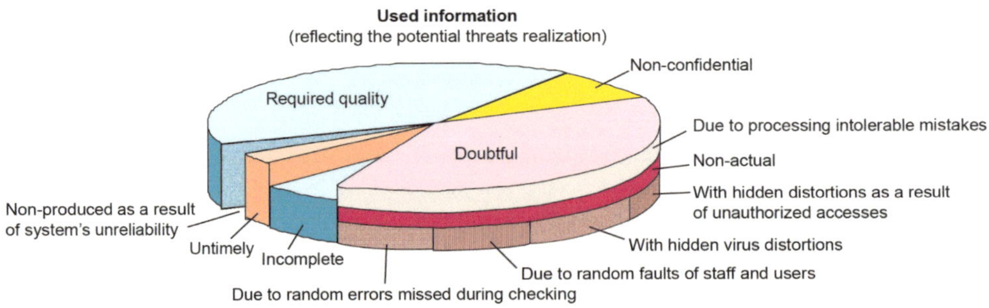

Figure 4.
The example of abstract information quality in the system.

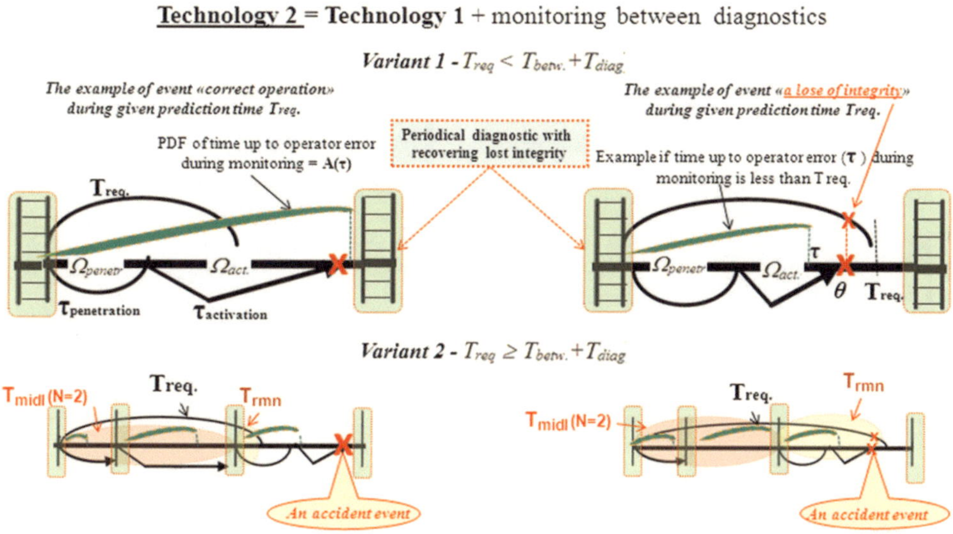

Figure 5.
Some accident events for technology 2, left—successful (correct) operation, right—a lose of integrity during given time T_{req}.

intelligence system or a system of support or their combination. The operator repairs the system after having detected the loss of integrity hazard—see **Figure 5**. Accordingly, the model assumption of operator's faultless action can do the full neutralization of the active hazard. Penetration is only possible if an operator makes an error. A dangerous influence occurs if the danger is activated before the next diagnostic. Otherwise, the source will be detected and neutralized during the next diagnostic.

The probability of a successful operation within a given period of time, i.e. the probability of "success" (P) may be estimated using the models presented in **Appendix B**. The risk to lose integrity (R) is an addition to 1 of the probability of successful operation, i.e. $R = 1 - P$ considering consequences. Damage from the consequences for the given period is taken into account as an additional characteristic of the calculated probability.

5.3 Algorithm to generate probabilistic models for complex system

The algorithm helps to implement concepts 1 and 5 for complex systems with parallel or serial structure [9–31] with the assumption of random variables independence. Let us consider the elementary structure for two independent parallel or series elements. Let us PDF of time between losses of the i-th element integrity is $B_i(t) = P(\tau_i \le t)$, then the time between successive integrity losses will be determined as follows:

1. for a system composed of serial independent elements is equal to the minimum of the two times τ_i: failure of 1st or 2nd elements. The PDF $B_{sys}(t)$ is defined by expression

$$B_{sys}(t) = P = [1 - B_1(t)]\bullet[1 - B_2(t)]; \tag{1}$$

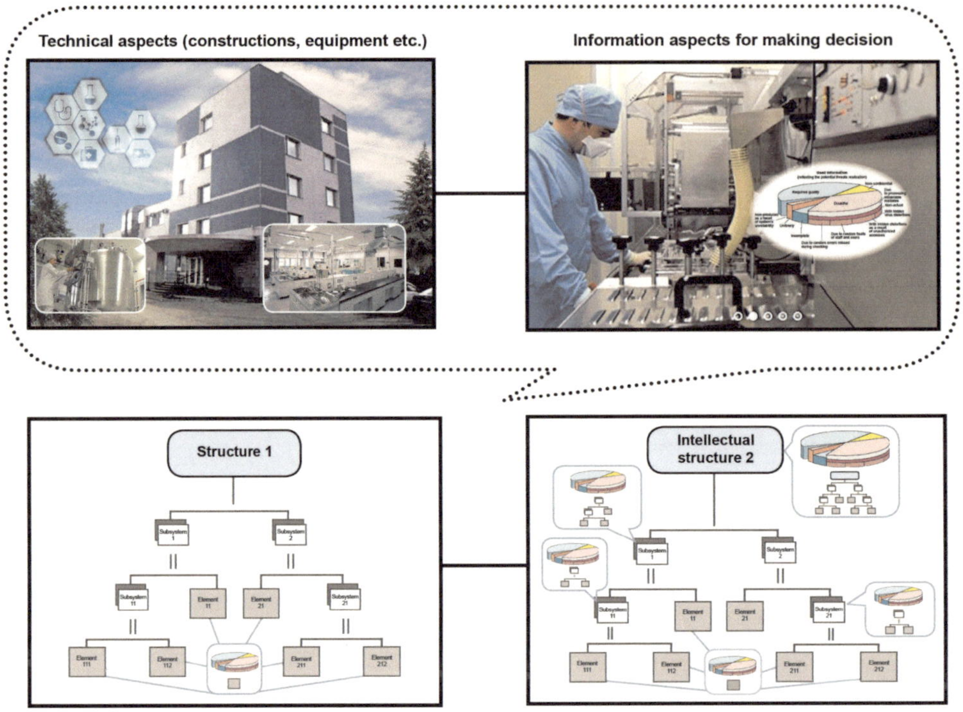

Figure 6.
An example of a complex system integrating two serial complex structures, which also are complex subsystems (abstraction).

2. for a system composed of parallel independent elements is equal to the maximum of the two timesτ_i, i.e. the system goes into the state of integrity loss when both elements lose integrity. The PDF $B_{sys}(t)$ is

$$B_{sys}(t) = P = [1 - B_1(t) \bullet B_2(t)]. \quad (2)$$

Applying expressions (1–2), the PDF of the time interval between successive losses of integrity for any complex system with parallel and/or serial structure and their combinations can be built. An example of a complex system integrating two serial complex subsystems is presented in **Figure 6**, see also Examples 2−4. For this system the following interpretation of elementary events is used: complex system integrating serial components "structures 1 and 2" is in the state of "successful operation" during a given period T_{req} if during this period component "structure 1" "AND" component "structure 2" are in the state of "successful operation". Note that both components are in their turn complex subsystems including subsystems and components, as well.

6. Risks prediction for standard processes

6.1 About standard processes

All actions in the timeline may be characterized as the performance of some system processes. The main system processes according to ISO/IEC/IEEE 15288 "System and software engineering—System life cycle processes" include 30 standard processes—agreement processes (acquisition and supply processes), organizational project-enabling processes (life cycle model management, infrastructure management, portfolio management, human resource management, quality management and knowledge management processes), technical management processes (project planning, project assessment and control, decision management, risk management, configuration management, information management, measurement and quality assurance processes), technical processes (business or mission analysis, stakeholder needs and requirements definition, system requirements definition, architecture definition, design definition, system analysis, implementation, integration, verification, transition, validation, operation, maintenance and disposal processes).

The focus on standard processes is justified by the fact that the life cycle of any complex system is woven from a variety of standard processes deployed in time, and for them, possible purposes, outcomes and typical actions are defined. Considering that for many critical systems, the potential damage and costs of eliminating the consequences in the conditions of heterogeneous threats can exceed the costs of preventive measures by an order of magnitude, it is necessary to find effective solutions to counter threats and ensure effective risk management for each of the processes performed. Despite many works on risk management for different areas theproblems of this chapter continue to be relevant (of course in practice developing new processes may be considered, not only from ISO/IEC/IEEE 15288 standpoint).

6.2 The example about input for probabilistic modelling

The proposed practical way to input forming helps to implement concept 6 for any monitored system (including real time system). For each critical parameter (for which prognostic estimations are needed to do proactive actions) the ranges of acceptable conditions can be established. The traced conditions of monitored parameters are data on a timeline. For example, the ranges of possible values of conditions may be established: "Working range within the norm", "Out of working range, but within the norm" and "Abnormality" for each traced separate critical parameter. If the parameter ranges of acceptable conditions are not established in explicit form, then for modelling purpose they may be impleaded and can be expressed in the form of average time value. These time values are

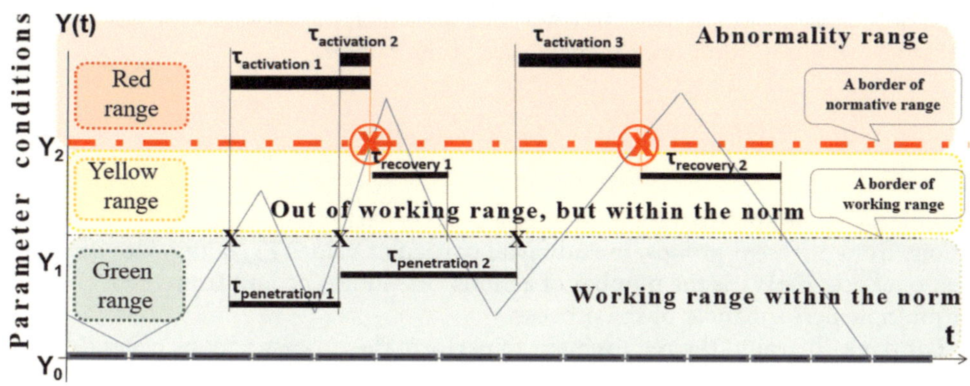

Figure 7.
An example of universal ranges for data about events and conditions. Note. In general case, the ranges may be established by subjective mode if a reasonable and objective one is impossible.

used as input for probabilistic models (see **Appendices A** and **B**). For example, for coal mine some of many dozens heterogeneous parameters may be compression, temperature, etc. It may be interpreted similarly by light signals "green", "yellow" and "red" [18, 25, 28–31]—see **Figure 7** and following Example 1.

6.3 The considerations

For the estimation of reliability of standard process performance, there may be two cases to estimate the probabilistic measure: the case of observed repeatability and the case of assumed repeatability of random events influencing reliability without the consideration of additional specific threats (for example, threats to operational information quality). For the estimation, the probabilistic measure repeatability of threats activation is assumed. For estimation of the assumption of independence of events connected with reliability of standard process performance and additional specific threats activations (for example, threats to information security) is used.

6.4 The case of the observed repeatability

The inputs for calculations use statistical data according to some observed repeatability. For standard process, the reliability of process performance and expected results in time are required. Failure to perform the necessary actions of the process is a threat of possible damage. From the different points of view, all varieties of the standard process can be divided into K groups, $K \geq 1$ (if necessary). Based on the use of statistical data, the probability $R_{\text{act } k}(T_k)$ of failure to perform the actions of the process for the k-th group for a given time $T_{\boldsymbol{k}}$ is proposed to be calculated by the formula:

$$R_{\text{act } k}(T_k) = G_{\text{failure } k}(T_{\boldsymbol{k}})/G_k(T_{\boldsymbol{k}}), \tag{3}$$

where $G_{\text{failure } k}(T_{\boldsymbol{k}})$,$G_k(T_{\boldsymbol{k}})$—are accordingly the number of cases of failures when performing the necessary actions of the process and the total quantity of necessary actions from the k-th group to be performed in a given time T_k.

The probability $R_{\text{rel}}(T)$ of failure to perform the necessary actions of a standard process without consideration of additional specific threats activations (for example, threats to operational information quality) is proposed to be estimated for the option when only those cases are taken into account for which the actions were not performed properly (they are the real cause of the failure)

$$R_{\text{rel}}(T) = 1 - \sum_{k=1}^{K} W_k[1 - R_{\text{act } k}(T_k)]\, Ind(\alpha_k)/\sum_{k=1}^{K} W_k, \tag{4}$$

where T is the specified total time for a process performance for the entire set of actions from different groups, including all particular values T_k, taking into account their overlaps; theW_k is the number of actions taken into account from the k-th group for multiple performances of the process.

For the k-th group, the requirement to perform the process actions using the indicator function $Ind_k(\alpha_k)$is taken into account

$$Ind(\alpha) = \begin{cases} 1, \textit{ifthe specified requirements} \wedge \textit{conditions are met, i.e.}\alpha \textit{ is true}, \\ 0, \textit{otherwise, i.e.if the condition } \alpha \textit{ is false}. \end{cases} \tag{5}$$

The condition α used in the indicator function is determined by the analysis of different specific conditions, proper to the process. It is to allow take into account the consequences associated with the failure of the process—see (3) and (4). Condition α_k means a set of conditions for all process actions required from the k-th group.

6.5 The case of the assumed repeatability

There may be recommended the models from Section 5 and **Appendices A** and **B**, which do not exhaust the whole set of possible probabilistic models.

6.6 About estimation of generalized measure

The generalized probability $R_{\text{gener}}(T)$ of failure to perform standard process considering additional specific threats $R_{\text{add}}(T)$ for the given period T may be calculated by the formula:

$$R_{\text{gener}}(T) = 1 - [1 - R_{\text{rel}}(T)] \cdot [1 - R_{\text{add}}(T)]. \quad (6)$$

Here the probabilistic measure $R_{\text{gener}}(T)$ of failure to perform reliable process considering specific threats are estimated according to proposition of section 5, subsections 6.1–6.5 and **Appendices A** and **B** considering the possible consequences.

6.7 Approach for risks integration from different processes

The integral risk of violation of the acceptable performance of standard processes set is proposed to be evaluated depending on the real or hypothetical initial data characterizing each process (see subsections 6.1–6.4), from the possible scenarios of using each process during the given prediction period. The prediction period can cover any period during the system life cycle, including at least 1 complete process of each of the types involved in the specified set of standard processes in the scenario under consideration. An example of the standard processes performed on the time axis is shown in **Figure 8**. In general, the scenario of using standard processes i_1, i_2, … , i_k is proposed to be determined by the real or hypothetical frequency of these

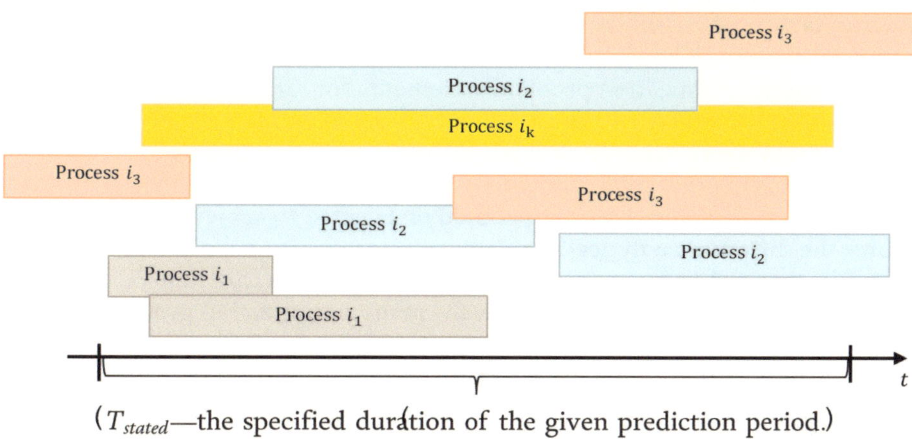

Figure 8.
An example of standard processes performed on the time axis.

processes (taking into account the typical actions performed at the same time that affect the time characteristics of the implementation of the processes). This approach allows us to take into account such opportunities when one process can be part of the work performed within other standard processes in the system lifecycle and, if necessary, includes other processes.

The integral risk of violation of the acceptable performance of standard processes set $R_{\int}(T_{stated})$ for given prediction period T_{stated} is proposed to be estimated by the formula

$$R_{\int}(T_{stated}) = 1 - \sum_{i=1}^{I} \lambda_i \{1 - R_i(T_{stated\ i}) \cdot [Ind(\alpha_i)]\} / \sum_{i=1}^{I} \lambda_i, \tag{7}$$

where λ_i is the expected frequency of execution of standard processes of the i-th type for the prediction period. If the duration of the executed process can go beyond the prediction period (which depends on the actions performed and their time characteristics), this frequency can be a fractional number that characterizes the number of executions of each type of process, greater than one;

$T_{stated\ i}$ – the expected period specified in the source data for modeling for the acceptable execution of standard type i process;

T_{stated} – the given prediction period that covers the duration of all the specified periods $T_{stated\ i}$ of each from standard processes involved in the scenario. The assumption about a partially completed process that can start at the end of the prediction period and not finish (if the total number of processes of each type is greater than one) can be satisfied by setting the fractional value λ_i.

At the same time, the criterion for meeting the requirements and conditions (α_i) for each type of process, including the requirements for acceptable risks and damages, is set using the indicator function (5).

Note. The expression (6) is a special case of expression (7).

The proposed in Section 6 models and methods are applicable for solving practical problems related to risk prediction and the justification of effective proactive measures to reduce risks or their prevention within acceptable limits.

7. Optimization of problem statements for rationale proactive actions

The proposed optimization problem statements for rationale actions help to implement concept 7. The matter is the metrics calculated in sections 5 and 6, in the models from **Appendixes A** and **B** depend on many parameters. The values of some parameters may be given and often variated within system life cycle. These values of some parameters may be specified or selected to achieve pragmatic goals and solve the different analytical problems of systems rational concept, development, utilization and support (described in Section 2). They are impacting the values of the estimated probabilistic risks. It means many optimization problems may be solved by rationale proactive actions connected with providing rational values of these parameters. For example, such parameters for optimization may be the duration of prediction period, parameters, impact on the information quality (see **Appendix A**), system structure, for the compound components: time between the end of diagnostic and the beginning of the next diagnostic, diagnostic time (see **Appendix B**) etc.

The proposed concepts 2–6 may be supported by the following typical optimization problem statements for various systems [9, 14, 28–31]:

1. on the stages of the system conceptual design, development, production and support: system parameters, software, technical and control measures (they are described by a set **Q** of parameters, which may be optimized) are the most rational for the given prediction period if the minimum of expenses *Z(Q)* can be reached

$$Z(\mathbf{Q}_{rational}) = \min_{\text{parameters of } \mathbf{Q}} Z(\mathbf{Q}), \tag{8}$$

   a. with limitations on probability of an admissible level of quality $P_{quality}(\mathbf{Q}) \geq P_{adm}$ and expenses for development, production, and support $C(\mathbf{Q}) \leq C_{adm}$ and under other development, operation or maintenance conditions; or

   b. with limitations on admissible risk to lose system integrity $R(\mathbf{Q}) \leq R_{adm}$ and expenses for development, production and support $C(\mathbf{Q}) \leq C_{adm}$ and under other development, operation or maintenance conditions; or

   c. with limitations based on a combination between 1a) and 1b);

2. utilization stage:

   - System parameters, software, technical and control measures (**Q**) are the most rational for the given period of system operation if the maximum of the probability of successful operation can be reached

$$P_{quality}(\mathbf{Q}_{rational}) = \max_{\text{parameters of } \mathbf{Q}} P_{quality}(\mathbf{Q}), \tag{9}$$

     a. with limitations on probability of an admissible level of quality $P_{quality}(\mathbf{Q}) \geq P_{adm}$ and expenses for operation $C(\mathbf{Q}) \leq C_{adm}$ and under other operation or maintenance conditions; or

     b. with limitations on the admissible risk to lose system integrity $R(\mathbf{Q}) \leq R_{adm}$ and expenses for operation $C(\mathbf{Q}) \leq C_{adm}$ and under other operation or maintenance conditions; or

     c. with limitations based on a combination between 2.1a)and 2.1b);

   - System parameters, software, technical and control measures *(Q)* are the most rational for the given period of system operation if the minimum of the risk to lose system integrity can be reached

$$R(\mathbf{Q}_{rational}) = \min_{\text{parameters of } \mathbf{Q}} R(\mathbf{Q}), \tag{10}$$

     a. with limitations on the quality $P_{quality}(\mathbf{Q}) \geq P_{adm}$ and expenses $C(\mathbf{Q}) \leq C_{adm}$ and under other operation or maintenance conditions; or

b. with limitations on the admissible risk to lose system integrity$R(\mathbf{Q}) \leq R_{adm}$ and expenses $C(\mathbf{Q}) \leq C_{adm}$ and under other operation or maintenance conditions; or

c. with limitations based on a combination between 2.2a) and 2.2b).

These statements may be retransformed into the other problems statements of expenses minimization for different limitations.

In system life cycle, there may be a combination of these formal statements.

Note. There may be another applicable variants of optimization.

8. Examples

The applications of the proposed approach cover: the analysis of the reliability of complex systems built from unreliable components; the estimation of the expected reliability and safety for complex constructions and intelligent manufacturing, the modelling of robotic and automated systems operating in cosmic space, the optimization of a centralized heat supply system, the analysis of the possibilities to keep "organism integrity" by continuous monitoring, the risk analysis during longtime grain storage, the control of timeliness, the completeness and validity of used information; the comparison between information security processes in networks; resources management and predicting quality for information systems operation; the estimation of human factor, the research of mutual monitoring operators actions for transport systems, rational dispatching of a sequence of heterogeneous repair works, the analysis of sea oil and gas systems vulnerabilities in conditions of different threats, the development of recommendations to reduce risks for the important land use planning (including Arctic region), the rationales of preventive measures by using "smart systems" etc.—see [9, 14–31]. Here the examples are intended to demonstrate some probabilistic risk prediction sectorial applications.

8.1 Example 1 of predicting the mean residual time before the next parameter abnormality

The example demonstrates system possibility on the base of solving the inverse problem by models described in subsection 5.2 and **Appendix B**. The research results are applied to rationale actions in real time for coal companies.

The conditions of parameters, traced by dispatcher intelligence centre, are data about a condition before and after the current moment of time. But always the scientifically justified predictions open new possibilities in the prevention of risks. With the use of predicted residual time, the responsible staff (mechanics, technologists, engineers, etc.) can determine the admissible time for rational changing of system operational regime to prevent negative events after expected parameter abnormality. For monitored critical parameters, the probabilistic approach to predict the mean residual time before the next parameter abnormality is proposed.

For every subsystem (element) monitored parameter, the ranges of possible values of conditions are established—see **Figures** 7 and **9**. The condition "Abnormality" means system (element) integrity loss (it may simply mean "system failure" that includes also "functional failure"). To prevent the possible cross-border abnormalities propagation, through the prediction of the residual time on the base of the data about parameter condition fluctuations. Given that the information quality also is estimated and provided (by using models from **Appendix A**).

The predicted residual time T_{resid} is the solution t_0 of the following equation:

$$R(T_{penetr}, t, T_{betw}, T_{diag}, T_{req.}) = R_{adm.}(T_{req}) \tag{11}$$

concerning of unknown parameter t, i.e. $T_{resid} = t_0$.

Here $R(T_{penetr}, t, T_{betw}, T_{diag}, T_{req.})$ is the risk to lose integrity, calculated according to the model of **Appendix B**. T_{penetr} is the probabilistic expectation of PDF $\Omega_{penetr}(\tau)$, defined by the transition statistical parameter from "green" to "yellow"—see **Figures 7** and **9**. The other parameters T_{betw} and T_{diag} in (11) are known—see **Appendix B**. The example explains a method to rationally estimate the value of prediction time T_{req}.

The properties of the function R(Tpenetr, t, Tbetw, Tdiag, Treq.) are the next:

- if t increases from 0 to ∞, for the same another parameter, the function $R(\ldots, t, \ldots)$ is monotonously decreasing from 1 to 0;

- if parameter T_{req} increases, from 0 to ∞ for the same another parameter, the function $R(\ldots, T_{req})$ is monotonously increasing from 0 to 1,i.e. for large T_{req}, the risk approaches to 1. It means that the such maximal x exists when t = x and $T_{req.}$ = x and $0 < R(T_{penetr}, x, T_{betw}, T_{diag}, x) < 1$, i.e. the mean residual time

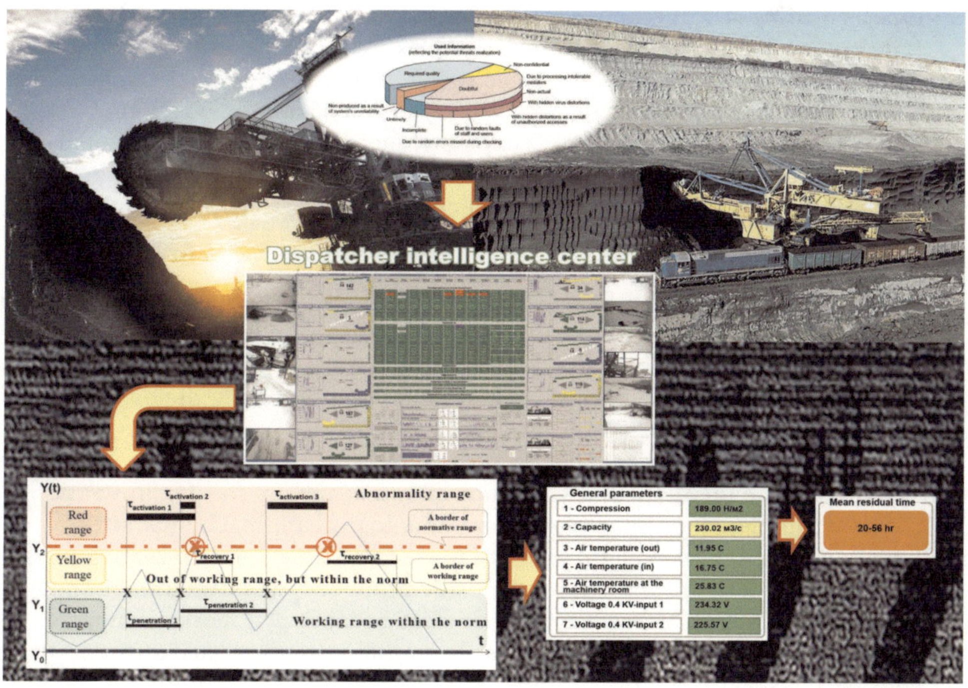

Figure 9.
Example of predicted residual time before the next parameter abnormality.

before the next abnormality is equal to "x" with the confidence level of the admissible risk $R(T_{penetr}, x, T_{betw}, T_{diag}, x)$. See details in [18].

The proposed ideas, probabilistic methods, models and justified normative requirements are implemented in Russia at the level of the national standard for system engineering—see for example GOST R 58494–2019 regarding the multifunctional safety systems of the coal mines (in the part of a remote monitoring system of critical industrial systems).

8.2 Examples related to socio-economic and pharmaceutical safety in a region

Examples 2–4 below demonstrate some analytical capabilities of the proposed approach for infrastructure management process related to socio-economic and pharmaceutical safety in a region of Russia. It concerns some problems in the creation and application of enterprise (S-N-T)-system – the manufacturer of pharmaceuticals denoted further as (S-N-T)-ESMP. Let the purposes of S-N-T-ESMP are to solve the following tasks:

- for the development of socio-economic infrastructure (tasks of the 1st type): ensuring the population with high-quality medications in a low-cost range (more than a hundred items); providing the emergence of new jobs; increasing tax revenues to the region; making profit from economic activities in the interests of strengthening and expanding business and stakeholders satisfaction;

- for the development of production and transport infrastructure (tasks of the 2nd type): ensuring strict compliance with the rules of production of good practice (GMP); development of a laboratory complex for ensuring and controlling the quality of products as part of microbiological, physical–chemical laboratories and air-liquid chromatography laboratories; expansion of the composition of manufacturers of substances and excipients, their suppliers and consumers of finished products; increasing the stability of the parameters of the production processes in order to ensure the reproducibility of the properties of finished medications;

- for the development of information and communication infrastructure (tasks of the 3rd type), providing the creation of an effective control system for ensuring the safety and quality, information security, integration of the enterprise into the state information system for monitoring the movement of medications.

In relation to the mentioned tasks, which allows achieving the demonstration goals of the examples, the application of methodological approach illustrates predicting on probability level: the risk of failure to reliable perform system infrastructure management process without consideration of specific abnormal impacts (see example 2); the risk of unacceptable damage because of abnormal impacts; the integral risk of failure to reliable perform system infrastructure management process considering specific abnormal impacts (see example 4). Assuming the commensurability of possible damages, a system analysis using probabilistic risk measures is carried out in the examples.

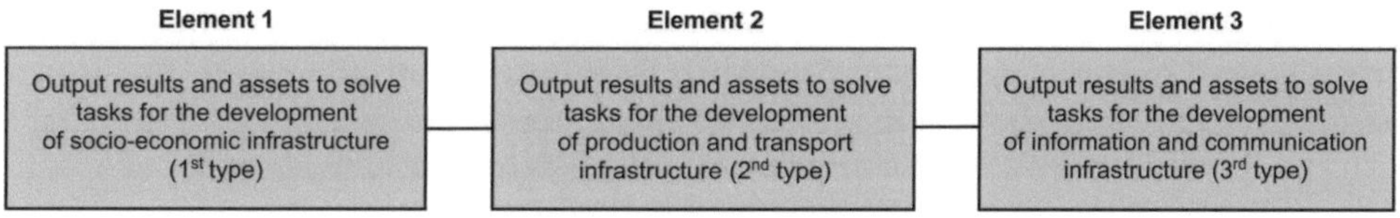

Figure 10.
The abstract complex structure for modelling (example 2).

Taking into account possible damages, the objectives of risk prediction are formulated as follows. In the conditions of existing uncertainty, to carry out: a quantitative predicting of the risks of failure to reliable perform system infrastructure management process without consideration of specific abnormal impacts; a quantitative predicting of the risks of unacceptable damage because of abnormal impacts on (S-N-T)-ESMP (both piecemeal for each type of infrastructure tasks and for the entire set of tasks); identification of critical factors affecting risks; determination of such a period in which guarantees of risks retention within admissible limits are maintained; a quantitative predicting of the integral risk of failure to reliable perform system infrastructure management process considering specific abnormal impacts.

Example 2. Here, the infrastructure model without consideration of specific abnormal impacts is focused on a set of output results and assets for solving tasks of the 1st, 2nd and 3rd types—see system structure in **Figure 10**. The following interpretation is applicable: During the given prediction period, the modeled complex structure is in an elementary state "the integrity of the infrastructure is maintained" if an implementation of the system infrastructure management process is reliable to solve the tasks "AND" for socio-economic, "AND" for production and transport "AND" for information and communication infrastructure. Many possible threats affecting the reliability of output results for each of the structural elements have been identified. Without delving into the numerous technical aspects of setting and solving the tasks of developing socio-economic, production and transport, information and communication infrastructure in a region, **Table 2** reflects hypothetical averaged input data for research by the models (see sections 5, 6 and **Appendix B** considering application of **Appendix A** models).

Input for every element (see model in Appendix B)	**Values**		
	for 1st element	**for 2nd element**	**for 3rd element**
σ—frequency of the occurrences of potential threats	2 times in a year	1 time in a year	1 time in a month
β—mean activation time of threats up to unacceptable damage	6 months	2 months	2 weeks
T_{betw} – time between the end of diagnostics and the beginning of the next diagnostics	1 week	1 week	1 hour
T_{diag} – diagnostics time of element integrity	1 hour	1 hour	1 minute
T_{recov} – recovery time after integrity violation	3 days	1 week	30 minutes
T – given prediction period	From 1 to 4 years		

Table 2.
Input for probabilistic modelling (example 2).

For modelling a period from 1 to 4 years was chosen because it is a typical period for short- and medium-term plans according to an infrastructure project. Analysis of the calculation results showed that in probabilistic terms the risk of failure to reliable perform system infrastructure management process without consideration of specific abnormal impacts for 2 years will be 0.282 totally for all elements (see **Figure 11**). In turn, for 1 year the risk will not fall below 0.150 (see **Figure 12**), and for 4 years with weekly diagnostics, the probabilities of "success" and "failure" will almost equal (0.51 vs. 0.49). In practice, such a level of risks is inadmissible, i.e. it is necessary to identify the critical factors (affecting risks) and effective ways to reduce risks.

Additional calculations have shown that one of the critical factors is the parameter "time between the end of diagnostics and the beginning of the next diagnostics" (that can also be called "Mean Time Before Diagnosis-MTBD" because of "diagnostics time of element integrity" is much less—see **Table 2**) for the 1st and 2nd elements (T_{betw}). Due to the management decision, expressed in changing the frequency of diagnostics from weekly to daily and the adoption of the appropriate measures to counter threats, with other conditions unchanged, it is possible to reduce risks several times. It is enough to compare the risks in **Figures 11** and **13**. About 2.1-fold reduction in risk has been achieved totally for all elements. That is, due to the most simply implemented

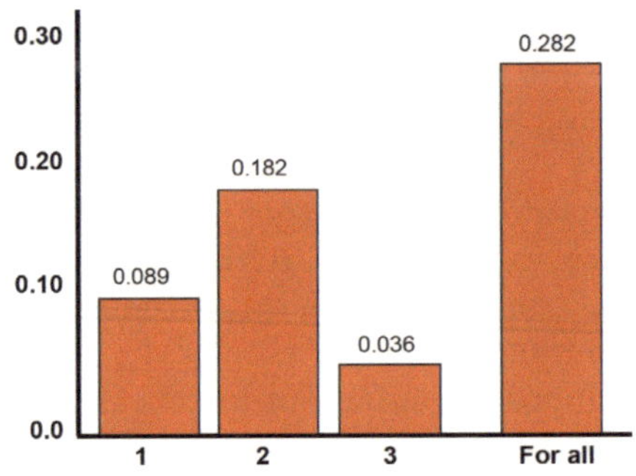

Figure 11.
The risks of failure to reliable perform system infrastructure management process without consideration of specific abnormal impacts on elements 1–3 for 2 years (for weekly diagnostics).

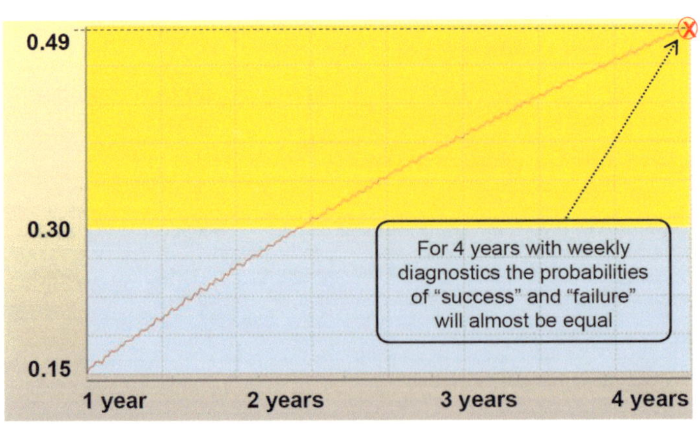

Figure 12.
The dependence of total risk of failure to reliable perform system infrastructure management process without consideration of specific abnormal impacts from duration of prediction period (for weekly diagnostics).

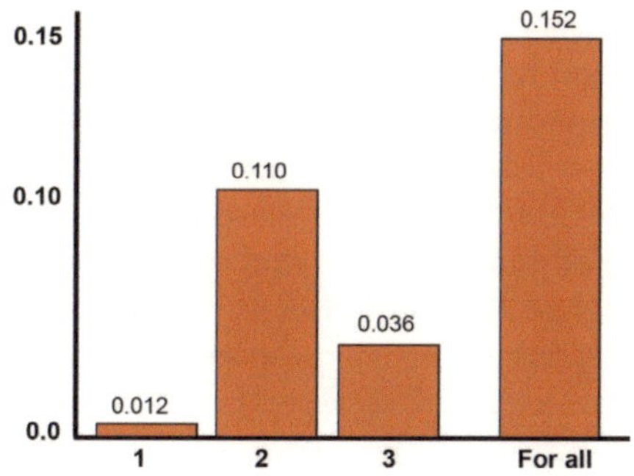

Figure 13.
The risks of failure to reliable perform system infrastructure management process without consideration of specific abnormal impacts on elements for 2 years.

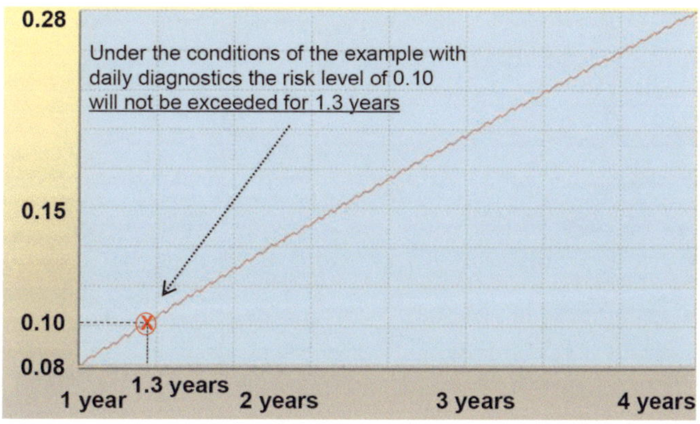

Figure 14.
The dependence of total risk of failure to reliable perform system infrastructure management process without consideration of specific abnormal impacts from duration of prediction period (for daily diagnostics).

organizational measures related to the introduction of more frequent diagnostics of work on the development of socio-economic and production and transport infrastructure, a significant risk reduction is achievable. This finding is the result of the used models. Despite the high value of this logical finding for example conditions, frequent diagnosis generates higher running costs and lower services supply capacity. Diagnosis costs money and time. It should be considered in other optimization problems (see Section 7).

For 1 year, the risk of failure of the infrastructure management process without considering the specific abnormal impacts will be about 0.08 (see **Figure 14**). In turn, as a result of the analysis of the risk dependence on the prediction period (from 1 to 4 years), it is additionally revealed that under the conditions of the example with daily diagnostics the risk level of 0.10 will not be exceeded for 1.3 years. Accordingly, and for infrastructure management process during development, focusing on admissible risk at the level of 0.10, in the conditions of the example, guarantees risks prevention within the admissible limits for about 1.3 years. Recommended measures to reduce risks are to increase the stability of the mechanical properties of the critical areas of structures, to timely carry on preventive and repair maintenance, to perform statistical analysis of emergency situations, and to predict critical unacceptable values of critical parameters inherent in the unacceptable risks.

Example 3. In contrast with Example 2, the model of specific abnormal impacts covers a set of actions related to the maintenance of buildings and constructions (element 1), ensuring the operation of engineering and technical systems (element 2), ensuring the operation of engineering networks (element 3), solving development problems of socio-economic infrastructure (element 4), production and transport infrastructure (element 5) and information and communication infrastructure (element 6)—see model on **Figure 15**. The following interpretation is applicable: during the given prediction period, the modeled complex structure is in an elementary state "the integrity of the system in case of abnormal impacts is maintained", if all the system elements are taken into account during the entire prediction period are in the state "the integrity of the system element in case of specific abnormal impacts is maintained".

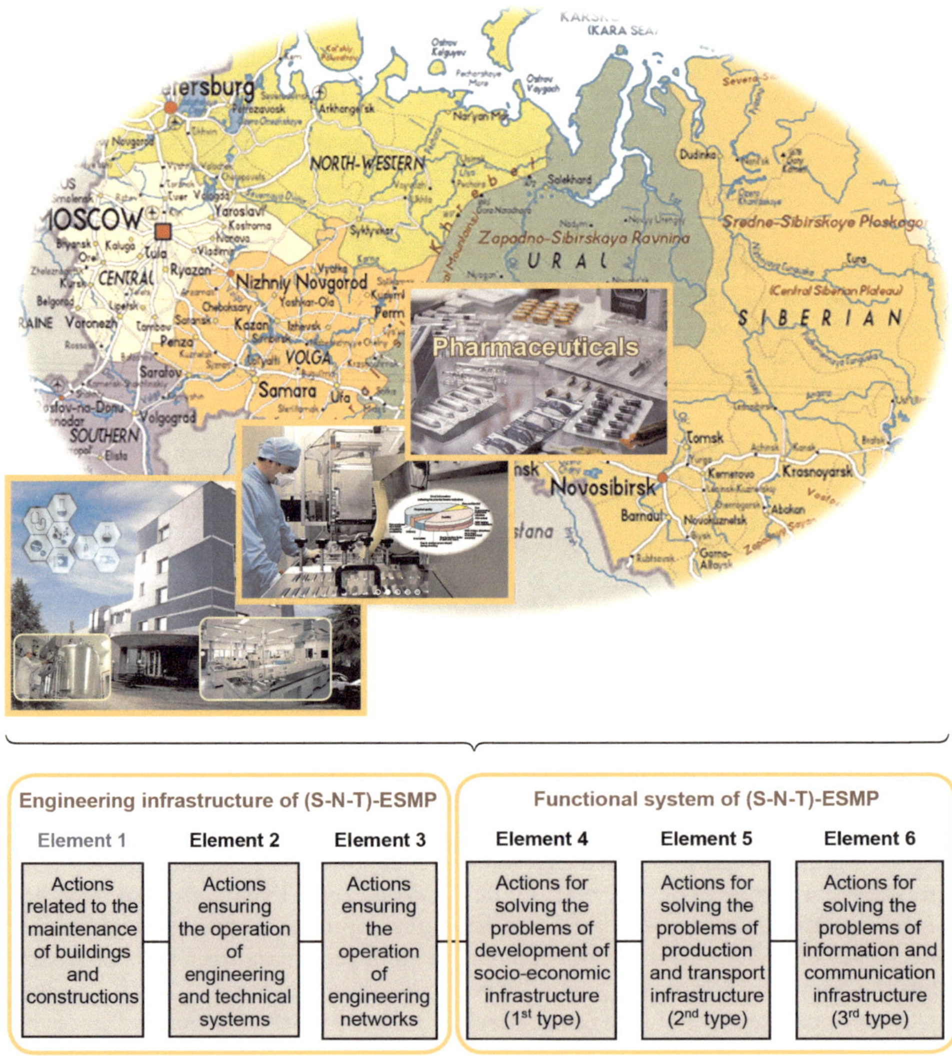

Figure 15.
The abstract complex structure for modelling specific abnormal impacts.

Input for every element (see model in [1–7])	Elements	Values
σ – frequency of the occurrences of potential threats	Element 1	4 times a year
	Element 2	2 times a year
	Element 3	1 time in a year
	Element 4	1 time in 2 years
	Element 5	1 time in 2 years
	Element 6	2 times a year
β – mean activation time of threats up to unacceptable damage	For all elements 1–6	1 month
T_{betw} – time between the end of diagnostics and the beginning of the next diagnostics	Element 1	24 hours
	Element 2	24 hours
	Element 3	24 hours
	Element 4	8 hours
	Element 5	8 hours
	Element 6	1 hour
T_{diag} – diagnostics time of element integrity	For all elements 1–6	30 seconds
T_{recov} – recovery time after integrity violation	For all elements 1–6	10 minutes
T – given prediction period	For all elements 1–6	From 1 to 4 years

Table 3.
Input for probabilistic modelling (example 3).

Without delving into the numerous technical, engineering and functional aspects of (S-N-T)-ESMP, **Table 3** reflects hypothetical averaged input data for research by models, described in sections 5–7, **Appendices A, B** and [1–7]. Input values for element 1 consider additional factors leading to degradation and destruction of techno-sphere systems (seismic, wind, snow, corrosion and other natural impacts). For element 6, proper impacts may be from "human factor" and/or from "cyber threats". For elements 2–5, input values have usual explanation.

Analysis of the results showed that in probabilistic terms the risk of unacceptable damage due to specific abnormal impacts for 2 years will be about 0.219 totally for all elements (see **Figure 16**). In turn, for the predict for 4 years with daily monitoring of the state of the engineering infrastructure of the (S-N-T)-ESMP (i.e. elements 1, 2, 3), the risk of unacceptable damage from specific impacts for all elements 1–6 will be about 0.39, and for the predict 1, this probability is about 0.12 (see **Figure 17**). In general, the results are comparable with the results of Example 2 (**Figures 18** and **19**).

Moreover, due to the management decision, expressed in changing the frequency of diagnostics from daily to 1 time every 8 hours it is possible to reduce total risk from 0.219 to 0.091 (see **Figure 18**). And owing to diagnostics every 8 hours the admissible risk level of 0.10 will not be exceeded about 2.3 years (see **Figure 19**).

Example 4. In continuation of Examples 2 and 3, the integral probability $\boldsymbol{R}_{\int}(\boldsymbol{T})$ of failure of the infrastructure management process considering specific system abnormal impacts for the prediction period T = 1 year is calculated using the recommendations of Section 6. It depends on probabilities $\boldsymbol{R}_{\mathbf{rel}}(\boldsymbol{T})$ and $\boldsymbol{R}_{\mathbf{add}}(\boldsymbol{T})$ – see formula (6). Considering that $\boldsymbol{R}_{\mathbf{rel}}(\mathbf{1year})$ = 0,08 and $\boldsymbol{R}_{\mathbf{add}}(\mathbf{1year})$ = 0,05,

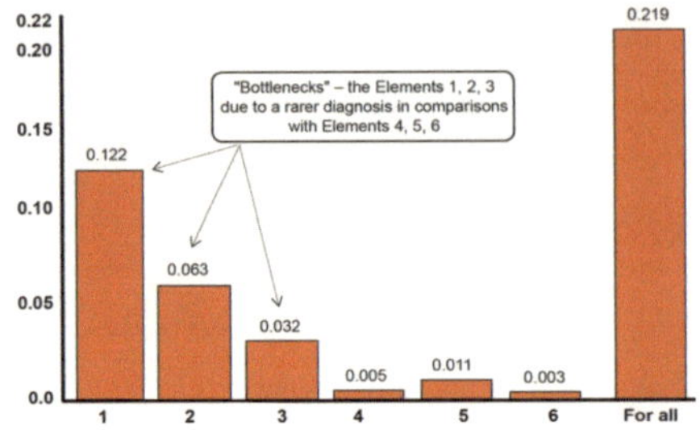

Figure 16.
The risks of unacceptable damage because of specific abnormal impacts on elements 1–6 for 2 years (for daily diagnostics).

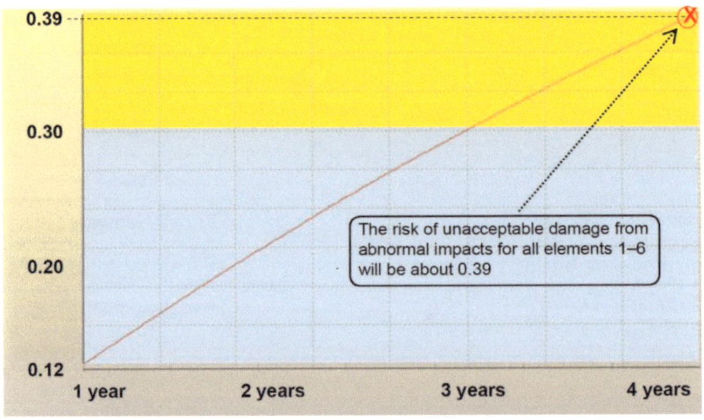

Figure 17.
The dependence of total risk of unacceptable damage because of specific abnormal impacts on elements 1–6 from duration of prediction period (for daily diagnostics).

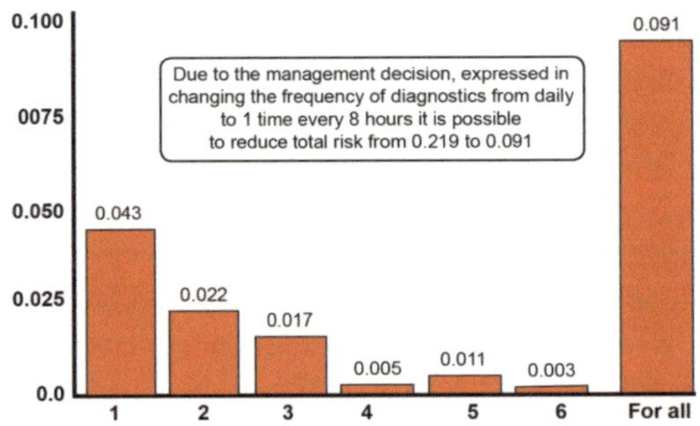

Figure 18.
The risks of unacceptable damage because of abnormal impacts on elements 1–6 for 2 years (for diagnostics every 8 hours).

$$R_{\int}(\mathbf{1year}) = 1\text{—}(1\text{—}0,\ 08) \cdot (1\text{—}0,\ 05) \approx 0{,}126.$$

Interpretation: the integral risk for the prediction period of 1 year is about 0.126 considering possible damage. In general, such risk is considered elevated. It can be considered acceptable only in exceptional cases when there are no real possibilities of any counteraction to threats. As measures to improve the process, additional control systems for damaging natural factors, emergency protection systems for

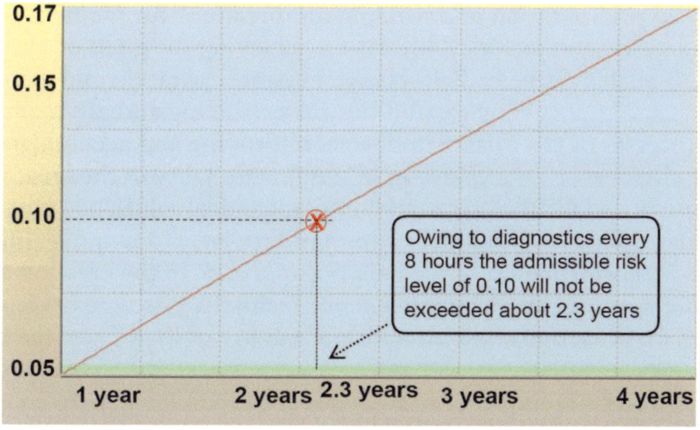

Figure 19.
The dependence of total risk of unacceptable damage because of abnormal impacts on elements 1−6 from duration of prediction period (for diagnostics every 8 hours).

techno-sphere systems, operators and personnel under extreme natural hazards, and measures to increase safety against specific system threats (the sources of specific abnormal impacts) can be used. Since such opportunities are far from being exhausted, an additional search for measures to reduce the integral risk is necessary. Decision-making on ways to reduce risks may be quantitatively justified using the proposed models and methods.

8.3 What about the possible pragmatic effects?

In general pragmatic effects are connected with achieving pragmatic goals (see Section 2). It may characterize the efficiency of the implementation of the state and/or corporate strategy in the economy, effects from improving the safety and sustainability of the region's development, from ensuring the protection of the population and territories from natural and man-made hazards, etc. For example, the authors of this chapter took part in the creation of the complex of supporting technogenic safety in the systems of oil and gas transportation and distribution and have been awarded for it by the award of the Government of the Russian Federation in the field of a science and technic. Through the Complex intellectual means, it is possible to detect remote-sensing technology: vibrations, fire, flood, unauthorized access, hurricane; and to recognize, identify and predict the development of extreme hazardous situations, and to make decisions in real-time. The applications of this Complex for 200 objects in several regions of Russia during the period 5 years have already provided economy about 8,5 Billion Roubles (reached at the expense of effective risks prediction and processes optimization [7]).

9. About directions for development

It is proposed to focus on scientific and technical efforts on the meta-level of system engineering, which allows, by universal probabilistic models, to set and analytically solve the problems of rational development and efficient operation of complex systems of various functionalities and purposes.

The proposed prioritization of development directions for predicting are: 1 – focusing on scientific and technical efforts on achieving the goals of ensuring the required safety, quality, balanced effects, sustainable operation and development of complex systems; 2 – providing capabilities for predicting and rational risk managing in standard processes of the system life cycle, improving and accumulating knowledge, patterns discovery; 3 – expansion of the functionality of the created models and methods, software and technological and methodological solutions (for predicting and rational risk managing) to all spheres of human activity, cross-application of knowledge bases; 4 – transformation of the existing approach to the creation and use of models and methods into artificial intelligence technology to support logical decision-making (based on proactive research with traceability of logic from the idea to the achieved effect).

The proposed steps to implement these directions are 1st step: from pragmatic filtering of information → to promising ideas and purposeful conceptions; 2nd step: from promising ideas and purposeful conceptions → to the formalization of uncertainties; 3rd step: from the formalization of uncertainties → to the knowledge of patterns and logical solutions; 4th step: from the knowledge of patterns and logical solutions → to rational risk management; 5th step: from rational risk management → to achieving the required safety, quality, balanced effects and sustainable operation and development.

The expected results will equally be understood at the level of probabilistic risk predictions, identically interpreted and comparable, the traceability of the effectiveness of scientific and technical system efforts from the conceptions to the results obtained will also be ensured. The purposeful aspirations "From uncertainties formalization – to sustainable harmony" (see Section 1) may be really supported.

10. Conclusion

On the generalizations of goals and objectives throughout the system's life cycle and existing approaches to risks prediction, there are proposed main goals of applying probabilistic methods. The goals of probabilistic concepts of risks prediction are connected with: an analysis of opportunities, achievable quality, safety and efficiency; a rationale for achieving desirable characteristics and acceptable conditions; an optimization of systems and processes; and finding and developing new ideas and concepts.

The enlarged classification of probabilistic methods for solving various objectives is explained for all stages of the system life cycle: concept, development, production, utilization, support and retirement.

The conceptual approach, proposed to risk prediction, covers social sphere (person, society, state and world community), natural sphere (earth and space) and techno-sphere (man-made infrastructures and life support facilities).

The essence of the proposed probabilistic concepts of risks prediction for the system is described on the level of probability distribution function. The described methods of risks prediction for complex systems include probabilistic models, methods for risks prediction and integration, optimization methods for rationale actions and examples for solving the problems of system analysis and rationale proactive actions in uncertain conditions. The achievable practical effects are explained.

The prioritization of development directions for risk prediction in standard system processes and targeted steps for their implementation are proposed. They support the purposeful aspirations "From uncertainties formalization – to sustainable harmony" in application to the life cycle of various systems.

Appendix A. The recommended models to predict information system operation quality

The proposed models are presented in **Table A.1**.

Appendix B. The models to predict risks for "Black box"

B.1. The model for technology 1 ("Black box") – see 5.2, [9, 14, 15].

Input:

$\Omega_{penetr}(t)$—is the PDF of time between neighboring influences for penetrating a danger source;

$\Omega_{activ}(t)$– is the PDF of activation time up to "accident event";

$T_{betw.}$—is time between the end of diagnostic and the beginning of the next diagnostic,

T_{diag}—is diagnostic time.

$R = 1 - P$ considering consequences.

Variant 1—$(T_{req.} < T_{betw.} + T_{diag})$:

$$P_{(1)}(T_{req.}) = 1 - \Omega_{penetr} * \Omega_{activ}(T_{req.}). \tag{12}$$

Variant 2—the assigned period $T_{req.}$ is more than or equals to an established period between neighboring diagnostics $(T_{req.} \geq T_{betw.} + T_{diag})$:

measure a)

$$P_{(2)}(T_{req.}) = N((T_{betw.} + T_{diag})/T_{req.}) \bullet P_{(1)}^{N}(T_{betw.} + T_{diag}) + (T_{rmn}/T_{req.}) \bullet P_{(1)}(T_{rmn}), \tag{13}$$

where $N = [T_{given}/(T_{betw.} + T_{diag})]$ is the integer part,

$$T_{rmn} = T_{given} - N(T_{betw.} + T_{diag});$$

measure b)

$$P_{(2)}(T_{req.}) = P_{(1)}^{N}(T_{betw.} + T_{diag}) \bullet P_{(1)}(T_{rmn}), \tag{14}$$

where the probability of success within the given time $P_{(1)}(T_{req.})$ is defined by (B.1).

B.2. The model for technology 2 ("Black box")—see 5.2, [9, 14, 15].

Input:

Additionally to Input for technology 1: $A(t)$—is the PDF of time from the last finish of diagnostic time up to the first operator error.

Evaluated measures:

Risk to lose system integrity (R). Probability of providing system integrity (P).

$R = 1 - P$ considering consequences.

Models. Input	Evaluated measures
The model of functions performance by a complex system in conditions of unreliability of its components. Input: $N(t)$—is the PDF of time between neighboring failures; $W(t)$—is the PDF of repair time; and $V(t)$—is the PDF of given time if this time is a random value	Probability P_{rel} of providing reliable function performance during given time $P_{rel} = \int_0^\infty \left\{ \int_t^\infty V(\tau - t) dN(\tau) \right\} dt / \int_0^\infty t d[N * W(t)]$, (A.1) *—is the convolution sign.
The model of calls processing for the different dispatcher technologies. Input for M/G/1/∞: λ_i– frequency of the i-th type calls for processing; β_i—mean processing time of the i-th type calls (without queue).	Probability $P_{tim.i}$ of well-timed processing of i-type calls during the required term $T_{req.i}$ $P_{tim.i} = P(t_{full.i} \le T_{req.i}) = \frac{\int_0^{\gamma_i^2 T_{req.i}/T_{full.i}} t^{\gamma_i^{-1}} e^{-t} dt}{\int_0^\infty t^{\gamma_i^{-1}} e^{-t} dt}$, (A.2) $\gamma_i = \frac{T_{full.i}}{\sqrt{T_{full.i2} - T_{full.i}^2}}$. Relative portion of all well-timed processed calls—S and relative portion of well-timed processed calls of those types for which the customer requirements are met—C: $S = \frac{\sum_{i=1}^I \lambda_i P_{tim.i}}{\sum_{i=1}^I \lambda_i}, C = \frac{\sum_{i=1}^I \lambda_i P_{tim.i} [Ind(\alpha_1) + Ind(\alpha_2)]}{\sum_{i=1}^I \lambda_i}$, $Ind(\alpha) = \begin{cases} 0, \text{ if } \alpha = true \\ 1, \text{ if } \alpha = false \end{cases}$, $\alpha_1 =$(criterion 1); $\alpha_2 =$(criterion 2). Criterion 1 is requirement to mean processing time $T_{full.i} \le T_{req.i}$, which means that i-type calls must be processed in time (in average), criterion 2 is requirement on the level of PDF $P_{tim.i} = P(t_{full.i} \le T_{req.i}) \ge P_{adm.i}$, which means hard processing in real time, $P_{adm.i}$—is admissible level for well-timed processing of i-type calls during the required term $T_{req.i}$.
The model of entering into system current data concerning new objects of application domain. Input: q_m—the probability that m new objects appear in random moment, intervals between these moments are exponentially distributed with parameter λ. $\Phi(z) = \sum_{m>0} q_m z^m$—is productive (generating) function; $B(t)$ —is the PDF of time for new information revealing and preparing, transfer and entering into database	Probability $P_{comp.}$ that system contains information about states of all real objects and coincides $P_{comp.} = exp\left\{ -\lambda \int_0^\infty [1 - \Phi(B(t))] \right\}$, (A.3)

The model of information gathering. Input: $C(t)$ is the PDF of time between essential changes of monitored object states, ξ_i—is the mean time for this PDF; $B(t)$ is the PDF of time for information gathering, preparing, transfer and entering into system; $Q(t)$ is the PDF of time between information updating, q is mean time; the mode D_1 of gathering: information is gathered in order "immediately after an essential object state change; the mode D_2 of gathering: information is gathered without any dependencies on changes in objects' current states (including regulated information gathering).	Probability P_{act} of information actuality on the moment of its use: 1) for the mode D_1 when information is gathered in order "immediately after an essential object state change: $P_{act} = \frac{1}{\xi_i}\int\limits_0^\infty B(t)[1 - C(t)]dt,$ (A.4) 2) for the mode D_2 when information is gathered without any dependencies on changes in objects' current states (periodical gathering) $P_{act} = \frac{1}{q}\int\limits_0^\infty \left\{[1 - Q(t)]\left[1 - \int\limits_0^\infty C(t+\tau)dB(\tau)\right]\right\}dt,$ (A.5)
The model of information analysis. Input: $T_{req.}$—assigned term for analysis; $N(t)$ – is the PDF of time between operator error ("false" instead of "true" (on time line), i.e. type I errors, η^{-1} is the mean time; $M(t)$ is the PDF of time between the neighboring errors in analyzed information (on timeline); $A(t)$ is the PDF of time between skipping an error (type II errors on timeline), T_{MTBF} is mean time; μ is the possible relative fraction of errors in information content (destined for problems of checking) or the possible relative fraction of information, which is essential for analysis (destined for problems of analysis); $T_{real} = V/\nu$—is the real time for complete information analysis; V—is a content of analyzed information; ν—is an analyzed speed; $T_{cont.}$—is time of continuous analyst's work. $T_{req.}$– is given term for analysis (deadline)	Probability P_{after} to be without errors after checking (or probability P_{after} of correct analysis): Variant 1—$(T_{real} \le T_{req.})$ and $(T_{real} \le T_{cont.})$: $P_{after(1)}(V, \mu, \nu, n, T_{MTBF}, T_{cont.}, T_{req.}) =$ $= [1 - \hat{N}(V/\nu)] \cdot \left\{\int\limits_0^{V/\nu} dA(\tau)[1 - M(V/\nu - \tau)] + \int\limits_{V/\nu}^\infty dA(t)\right\}$ (A.6) Variant 2—$(T_{real} \le T_{req.})$ and $(T_{real} > T_{cont.})$: $P_{after(2)} = \{P_{after(1)}(V_{part(2)}, \mu, \nu, \eta, T_{MTBF}, T_{cont.}, \tau_{part(2)})\}^N,$ (A.7) $N = V/(\nu \cdot T_{cont.}), V_{part(2)} = V/N, \tau_{part(2)} = T_{req.}/N.$ Variant 3—$(T_{real} > T_{req.})$ and $(T_{real} \le T_{cont.})$: $P_{after(3)} = (V_{part(3)}/V) \cdot P_{after(1)} \cdot V_{part(3)}, \mu, \nu, \eta, T_{MTBF}, T_{cont.},$ $+[(V - V_{part(3)})/V] \cdot P_{without},$ (A.8) where $V_{part(3)} = \nu \cdot T_{req.}, P_{without} = e^{-\mu \cdot (V - V_{part(3)})}.$ Variant 4—$(T_{real} > T_{req.})$ and $T_{real} > T_{cont.}$. $P_{after} =$ $= \begin{cases} [V_{part(4)}/V] \cdot P_{after(1)} \cdot (V_{part(4)}, \mu, \nu, \eta, T_{MTBF}, T_{cont.}, T_{req.}) + \\ +\left[(V - V_{part(4)})/V \cdot e^{-\mu \cdot (V - V_{part(4)})}\right), \text{if} T_{req.} \le T_{cont.}; \end{cases}$ $[V_{part(4)}/V] \cdot \{P_{after(1)} \cdot (V_{part(4.2)}, \mu, \nu, \eta, T_{MTBF}, T_{cont.}, \tau_{part(4.2)})\}^N +$ $+\left[(V - V_{part(4)})/V\right] \cdot e^{-\mu \cdot (V - V_{part(4)})}, \text{if} T_{req.} > T_{cont.},$ (A.9)

The model of authorized access to system resources during objective period. Input: M is the conditional quantity of security barriers in counteraction to unauthorized access; $F_m(t)$ is the PDF of time between changes of m-th barrier parameters, f_m is mean time for this PDF; $U_m(t)$ is the PDF of parameters decoding time of the m-th security system barrier, $H(t)$—is the PDF of objective period, when resources value is high. If the mean objective period, when resources value is high $\rightarrow \infty$, this he model is transformed into the model of an authorized access to system resources (see below)	Probability P_{value} of system protection against unauthorized access during objective period $P_{value} = 1 - \prod_{m=1}^{M} P_{over.m}$, (A.10) where $P_{over.m}$—is the risk of overcoming the m-th barrier by violator during objective period when resources value is high, $P_{over} = \frac{1}{f_m}\int_0^{\infty} dt \int_t^{\infty} dF_m(\tau) \int_0^{t} dU_m(\theta)[1 - H(\theta)]$.
The model of dangerous influences on a protected system. Input: $\Omega_{penetr}(t)$—is the PDF of time between neighboring influences for penetrating danger source; $\Omega_{activ}(t)$—is the PDF of activation time of penetrated danger source; $T_{req.}$—is the required period of permanent secure system operation; $T_{betw.}$ is time between the end of diagnostic and the beginning of the next diagnostic, $T_{diag.}$—is diagnostic time; $A(t)$—is the PDF of time from the last finish of diagnostic time up to the first operator error.	Probability P_{infl} of faultless (correct) operation during given time: Variant 1—$(T_{req.} < T_{betw.} + T_{diag})$: $P_{(1)}(T_{req.}) = 1 - \int_0^{T_{req.}} dA(\tau) \int_{\tau}^{T_{req.}} d\Omega_{penetr} {}^{*}\Omega_{activ}(\theta)$. (A.11) Variant 2—$(T_{req.} \geq T_{betw.} + T_{diag})$: measure a) $P_{(2)}(T_{req.}) = N((T_{betw.} + T_{diag})/T_{req.}) \cdot P_{(1)}^{N}(T_{betw.} + T_{diag}) +$ $+(T_{rmn}/T_{req.}) \cdot P_{(1)}(T_{rmn})$, (A.12) measure b) $P_{(2)}(T_{req.}) = P_{(1)}^{N}(T_{betw.} + T_{diag}) \cdot P_{(1)}(T_{rmn})$, (A.13) where N is the same and the probability of success within the given time $P_{(1)}(T_{req.})$, which is defined by (A.11)
The model of authorized access to system resources. Input (for estimation of confidentiality): M is the conditional quantity of a barriers against an unauthorized access; $F_m(t)$ is the PDF of time between changes of the m-th barrier parameters, f_m is the mean time for this PDF; $U_m(t)$ is the PDF of parameters decoding time of the m-th security system barrier	Probability P_{prot} of system protection against unauthorized access: $P_{prot} = 1 - \prod_{m=1}^{M} P_{over.m}$, (A.14) where $P_{over.m}$—is the probability of overcoming the m-th barrier by violator, $P_{over.m} = \frac{1}{f_m}\int_0^{\infty}[1 - F_m(t)] \bullet U_m(t)dt$. (A.15)

Note. The final clear analytical formulas are received by Lebesque-integration of (A.1), (A.3)–(A.6), (A.10), (A.11), and (A.15).

Table A.1.
The proposed models (for the details—See [7–9, 14, 15]).

For variant 1 $\left(T_{req.} < T_{betw.} + T_{diag}\right)$: see (A.11).

For variant 2 $\left(T_{req.} \geq T_{betw.} + T_{diag}\right)$: see (A.12), (A.13), and the same (B.2), (B.3).

Evaluated measures:

Risk to lose system integrity (R). Probability of providing system integrity (P).

Acknowledgements

The authors would like to thank the Russian Academy of Sciences for supporting the work presented in this Chapter.

The authors much obliged to Dr. Mohamed Eid, President of ESReDA (European Safety, Reliability & Data Association) for meaningful advices regarding Chapter.

And also thanks a lot to Ms. Olga Kostogryzova for some artistic Figures design.

Author details

Andrey Kostogryzov[1]*, Nikolay Makhutov[2], Andrey Nistratov[1] and Georgy Reznikov[3]

1 Federal Research Center "Computer Science and Control" of the Russian Academy of Sciences, Moscow, Russia

2 The A.A. Blagonravov Institute for Machine Sciences of the Russian Academy of Sciences, Moscow, Russia

3 Company "Regional Engineering Consulting Firm" Ltd., Samara, Russia

*Address all correspondence to: akostogr@gmail.com

References

[1] Feller W. An Introduction to Probability Theory and Its Applications. New York: Jhon Wiley & Sons; 1971;**2**

[2] Martin J. System Analysis for Data Transmission. Englewood Cliffs; New Jersey: Prentice Hall, Inc; 1972

[3] Gnedenko BV et al. Priority Queueing Systems. Moscow: MSU; 1973

[4] Kleinrock L. Queueing Systems, V.2: Computer Applications. New York: John Wiley & Sons; 1976

[5] Matweev VF, Ushakov VG. Queuing Systems. Moscow: MSU; 1984

[6] Kostogryzov AI, Petuhov AV, Scherbina AM. Foundations of Evaluation, Providing and Increasing Output Information Quality for Automatized System. Moscow: Armament. Policy. Conversion; 1994

[7] Security of Russia. Legal, Social& Economic and Scientific&Engineering Aspects. The Scientific Foundations of Technogenic Safety. Under the editorship of Makhutov N.A. Moscow: Znanie. Volumes 1–63; 1998-2021

[8] Kostogryzov AI. Software Tools Complex for Evaluation of Information Systems Operation Quality (CEISOQ). In: Proceedings of the 34-th Annual Event of the Government Electronics and Information Association (GEIA), Engineering and Technical Management Symposium. Dallas; 2000. pp. 63-70

[9] Kostogryzov A, Nistratov G. Standardization, Probabilistic Modelling, Rational Management and Certification in the Field of System and Sengineering (80 Standards, 100 Probabilistic Models, 35 Software Tools, More than 50 Practical Examples). Moscow: Armament. Policy. Conversion; 2005

[10] Zio En. An Introduction to the Basics of Reliability and Risk Analysis. Singapore: World Scientific Publishing Co.Pte. Ltd; 2006

[11] Makhutov NA. Strength and Safety. Fundamental and Applied Research. Novosibirsk: Nauka; 2008

[12] Kolowrocki K, Soszynska-Budny J. Reliability and Safety of Complex Technical Systems and Processes. London: Springer-Verlag; 2011

[13] Eid M, Rosato V. Critical Infrastructure Disruption Scenarios Analyses via Simulation. Managing the Complexity of Critical Infrastructures. A Modelling and Simulation Approach. New York: Springer Open; 2016:43-62

[14] Kostogryzov AI, Stepanov PV. Innovative Management of Quality and Risks in Systems Life Cycle. Moscow: Armament. Policy. Conversion; 2008

[15] Kostogryzov A, Nistratov G, Nistratov A. Some Applicable Methods to Analyze and Optimize System Processes in Quality Management. Total Quality Management and Six Sigma. Rijeka, Croatia: InTech; 2012:127-196

[16] Kostogryzov A, Nistratov G, Nistratov A. The innovative probability models and software technologies of risks prediction for systems operating in various fields. International Journal of Engineering and Innovative Technology (IJEIT). 2013;**3**(3):146-155

[17] Grigoriev L, Guseinov C, Kershenbaum V, Kostogryzov A. The methodological approach, based on the risks analysis and optimization, to research variants for developing hydrocarbon deposits of Arctic regions.

Journal of Polish Safety and Reliability Association. 2014;**5**:1-2

[18] Artemyev V, Kostogryzov A, Rudenko J, Kurpatov O, Nistratov G, Nistratov A. Probabilistic methods of estimating the mean residual time before the next parameters abnormalities for monitored critical systems. In: Proceedings of the 2nd International Conference on System Reliability and Safety (ICSRS). Milan, Italy; 2017. pp. 368-373

[19] Kostogryzov A, Stepanov P, Nistratov A, Atakishchev O. About Probabilistic Risks Analysis During Longtime Grain Storage. Proceedings of the 2nd Internationale Conference on the Social Science and Teaching Research (ACSS-SSTR), Volume 18 of Advances in Social and Behavioral Science. Singapore: Singapore Management and Sports Science Institute, PTE.Ltd; 2017: 3-8

[20] Kostogryzov A, Stepanov P, Nistratov A, Nistratov G, Klimov S, Grigoriev L. The method of rational dispatching a sequence of heterogeneous repair works. Energetica. 2017;**63**(4): 154-162

[21] Kostogryzov A, Stepanov P, Grigoriev L, Atakishchev O, Nistratov A, Nistratov G. Improvement of existing risks control concept for complex systems by the automatic combination and generation of probabilistic models and forming the storehouse of risks predictions knowledge. In: Proceedings of the 2nd International Conference on Applied Mathematics, Simulation and Modelling (AMSM). Phuket, Thailand: DEStech Publications; 2017. pp. 279-283

[22] Kostogryzov A, Panov V, Stepanov P, Grigoriev L, Nistratov A, Nistratov G. Optimization of sequence of performing heterogeneous repair work for transport systems by criteria of timeliness. In: Proceedings of the 4th International Conference on Transportation Information and Safety (ICTIS). Canada; 2017. pp. 872-876

[23] Kostogryzov A, Grigoriev L, Golovin S, Nistratov A, Nistratov G, Klimov S. Probabilistic modeling of robotic and automated systems operating in cosmic space. In: Proceedings of the International Conference on Communication, Network and Artificial Intelligence (CNAI). Beijing, China: DEStech Publications; 2018. pp. 298-303

[24] Kostogryzov A, Grigoriev L, Kanygin P, Golovin S, Nistratov A, Nistratov G. The experience of probabilistic modeling and optimization of a centralized heat supply system which is an object for modernization. In: International Conference on Physics, Computing and Probabilistic Modeling (PCMM). Shanghai: DEStech Publications, Inc; 2018. pp. 93-97

[25] Artemyev V, Rudenko J, Nistratov G. Probabilistic modeling in system engineering. Chapter 2. Probabilistic methods and technologies of risks prediction and rationale of preventive measures by using "smart systems". Applications to coal branch for increasing Industrial safety of enterprises. IntechOpen. 2018:23-51

[26] Kershenbaum V, Grigoriev L, Nistratov A. Probabilistic modeling in system engineering. Chapter 3. Probabilistic modeling processes for oil and gas systems. IntechOpen. 2018:55-79

[27] Kostogryzov A, Nistratov A, Nistratov G, Atakishchev O, Golovin S, Grigoriev L. The probabilistic analysis of the possibilities to keep "organism integrity" by continuous monitoring. In: Proceedings of the International

Conference on Mathematics, Modelling, Simulation and Algorithms (MMSA). Chengdu, China: Atlantis Press; 2018. pp. 432-435

[28] Kostogryzov A, Korolev V. Probability, combinatorics and control. Chapter 1. Probabilistic methods for cognitive solving problems of artificial intelligence systems operating in specific conditions of uncertainties. IntechOpen. 2020:3-34

[29] Kostogryzov A, Kanygin P, Nistratov A. Probabilistic comparisons of systems operation quality for uncertainty conditions. RTA&A. 2020; **15**:63-73

[30] Kostogryzov A, Nistratov A, Nistratov G. Analytical risks prediction. Rationale of system preventive measures for solving quality and safety problems. In: Sukhomlin V, Zubareva E, editors. Modern Information Technology and IT Education. SITITO 2018. Communications in Computer and Information Science. Cham: Springer; 2020. pp. 352-364. DOI: 10.1007/978-3-030-46895-8_27

[31] Kostogryzov A, Nistratov A. Probabilistic methods of risk predictions and their pragmatic applications in life cycle of complex systems. In: Safety and Reliability of Systems and Processes. Poland: Gdynia Maritime University; 2020. pp. 153-174

[32] Moskvichev VV, Makhutov NA, Shokin YM, Gadenin MM. Applied Problems of Structural Strength and Mechanics of Destruction of Technical Systems. Novosibirsk: Nauka; 2021

[33] Gneiting T, Balabdaoui F, Raferty AE. Probabilistic forecasts, calibration and sharpness. Journal of the Royal Statistical Society, Series B (Statistical Methodology). 2007;**68**:243-268

[34] Kostogryzov A, Nistratov A, Zubarev I, Stepanov P, Grigoriev L. About accuracy of risks prediction and importance of increasing adequacy of used probabilistic models. Journal of Polish Safety and Reliability Association. Summer Safety and Reliability Seminars. 2015;**6**:71-80

Section 5

Approaches and New Methods

Chapter 5

A New Approach of Power Transformations in Functional Non-Parametric Temperature Time Series

Haithem Taha Mohammed Ali
and Sameera Abdulsalam Othman

Abstract

In nonparametric analyses, many authors indicate that the kernel density functions work well when the variable is close to the Gaussian shape. This chapter interest is on the improvement the forecastability of the functional nonparametric time series by using a new approach of the parametric power transformation. The choice of the power parameter in this approach is based on minimizing the mean integrated square error of kernel estimation. Many authors have used this criterion in estimating density under the assumption that the original data follow a known probability distribution. In this chapter, the authors assumed that the original data were of unknown distribution and set the theoretical framework to derive a criterion for estimating the power parameter and proposed an application algorithm in two-time series of temperature monthly averages.

Keywords: functional non-parametric time series, power transformation, Kernel density function, Mean Integrated Square Error

1. Introduction

One of the most common approaches for studying forecasting models is the Nonparametric functional regression method, which has been successfully applied in time series analysis. In this chapter, a new approach of power transformation is proposed to improve time series prediction when using functional nonparametric techniques. Although the nonparametric regression estimation under dependence is a useful tool for forecasting in time series [1], the functional and nonparametric approaches does not work well in certain circumstances.

Regarding the functional approach, The functional data (FD) analysis treats with the observations as a functions [2] without the need for fully parametric and non-parametric modeling conditions. In other words, FD analysis reduces the size of the data by clarifying the correlations between a large number of variables by a small number of factors or functions [3]. This transformation of the data structure into a linear combination of a few functions (curves) is equivalent to structural regression models. A number of useful semi-metrics families can be used to measure the proximities between the curves of the functional variables. One of these ways, for example, the Functional Principle Components Analysis (FPCA) [4, 5]. In some data sets with the time dependence of observations, FPCA may lead to weak estimates and that this problem may be exacerbated in some time series data sets especially those characterized by the presence of seasonal changes [6] (See also [7] who pointed out that the standard PCA may not be the suitable technique to apply when the data distribution is skewed or there are outliers).

As for the nonparametric approach to estimating kernel density functions (KDF) or predictions in regression and time series models, although this approach is a Distribution-Free method, the symmetricity of the data is an important issue in order to obtain efficient estimators [8, 9].

As for time series and the goal of improving forecastability, it is known that the time series data sets in practical applications are rarely adapted for statistical analysis due to their instability in variance, trend, and seasonal variations [10].

Based on the aforementioned requirements of importance that precede the analysis and inference in the functional nonparametric time series analysis, it can be said that the power transformation (PT) provide a novel corrective framework of the predictive modeling.

The rest of the chapter is organized as follows: The second section includes some explanations and clarifications of some traditional approaches of PT and their uses in KDF. In the third section, the authors present their proposal which contains a new approach for Transforming of KDS. Section four includes the algorithm for applying the proposed method. While the fifth section includes an applications of the proposed method to two temperature time series datasets. Finally, the sixth section included the conclusions and some future recommendations.

2. The traditional approaches of transformations in KDF

There is a long tradition of applying PT models in statistical applications. In 1952, Finney [11] used the PT model $\boldsymbol{\Psi}(\boldsymbol{Z}) = \boldsymbol{Z}^{\lambda}$ when $\boldsymbol{\lambda} \neq \mathbf{0}$ and $\boldsymbol{\Psi}(\boldsymbol{Z}) = \boldsymbol{log}\,(\boldsymbol{z})$ when $\boldsymbol{\lambda} = \mathbf{0}$, where Z represents the original dose variable in the biological assay. The purpose of using transformation in dose response relationship was to achieve the monotonous and linear characteristics for the Intrinsically nonlinear models. In 1964, Box and Cox [12] proposed the following general class of transformation of the response variable in the multiple linear regression model,

$$\boldsymbol{\Psi}(\boldsymbol{Z}) = \begin{cases} \dfrac{\boldsymbol{Z}^{\lambda} - \mathbf{1}}{\boldsymbol{\lambda}} & \boldsymbol{if}\ \boldsymbol{\lambda} \neq \mathbf{0} \\ \boldsymbol{log}\,(\boldsymbol{z}) & \boldsymbol{if}\ \boldsymbol{\lambda} = \mathbf{0} \end{cases} \tag{1}$$

to achieve a linear relationship with normality errors. In 1977, Tukey [13] describes an orderly way of re-expressing variables using the following model in order to preserve the order of the variable after the PT is used,

$$\boldsymbol{\Psi}(\boldsymbol{Z}) = \begin{cases} \boldsymbol{Z}^{\lambda}\ \boldsymbol{if}\ \boldsymbol{\lambda} > \mathbf{0} \\ \boldsymbol{log}\,(\boldsymbol{z})\ \boldsymbol{if}\ \boldsymbol{\lambda} = \mathbf{0} \\ -\boldsymbol{Z}^{\lambda}\ \boldsymbol{if}\ \boldsymbol{\lambda} < \mathbf{0} \end{cases} \tag{2}$$

to make the relationship as close to a straight line as possible. As for the nonparametric estimates, many authors refer to the usefulness of PT in reducing the bias of KDF when the data is clearly skewed or heavy tailed [14], (For more details see, [15–17]).

Regarding the transformation parameter estimation issues, most transformation methodologies have a common analytical path, which is the choice of the PT model and proposing an algorithm for estimating the power parameters in parallel with the mechanisms of estimating traditional model parameters. As for the approach of transforming the probability density function (PDF) of all or some model variables before proposing an algorithm estimation, there are at least two common methodologies of data transformation. The first in chronological order is the Box Cox transformation (BCT) methodology of data transforming to normality of response variable in parametric multiple regression models [12]. The common decision rule for selecting power parameter estimator in this approach is the maximization of log likelihood function of the PDF of original data. In some cases, the Bayesian estimating method is used and many other methods included in the subject literature can also be used to choose the transformation parameter. The second methodology is proposed by Wand, Marron and Ruppert in 1991 [8] to transform the KDF to a symmetrical shape in density function. In this methodology, the decision rule for selecting optimal estimator of density power parameter is the minimization of the Mean Integrated Square Error (MISE) of KDF estimator. Both transformation ways are used the distribution approach of transformed data and therefor defining the original data distribution as a "back-transformed" of change-of-variable technique. Mathematically, in the case of the univariate random variable $\boldsymbol{Z}$, Box and Cox [12] assumes that there exist a parametric PT function $\psi(.)$ of the random variable $\boldsymbol{Z}$ such that $\psi(z) = \boldsymbol{Z}^{(\lambda)} \sim \boldsymbol{N}(\mu, \sigma^2)$ under the assumption that the original data is of unknown distribution. Therefore, the PDF of the original variable is given by,

$$f_Z(z) = f_{\psi(z)}(\psi(z_i); \mu, \sigma^2) . \left| \frac{d\psi(z)}{dz_i} \right| \tag{3}$$

While the second methodology [8], Wand, Marron, and Ruppert assumes the estimated KDF of the transformed variable $\psi(z)$ that is close to the symmetrical shape is given by,

$$f_{\psi(z)}(\psi(z); \lambda) = f_Z\{\psi^{-1}(z)\}(\psi^{-1})'(\psi(z)) \tag{4}$$

Such that the estimated KDF of the original variable $\boldsymbol{Z}$ is the back-transform of (Eq. (4)) and given by,

$$\hat{f}_z(z; h, \lambda) = n^{-1} \sum \psi'(z) k_h\{\psi(z) - \psi(Z_i)\} \tag{5}$$

Where h is the bandwidth and the kernel K is a density. In brief terms, the first methodology aims in the parametric models to improve the efficiency of the statistical inference based on the data normality, and the second methodology aims to improve the kernel estimator at least on the basis of symmetrical data. And in the same context, the literatures recommend the use of transformations as long as they can improve interpretation of effect sizes between variables [13] or given the fact that model parameters are not easily interpreted in terms of the original response [14], (For more details, see [15–17]).

Now, assuming that $U = \psi(z)$, The optimum value of the PT parameter λ is the one that corresponds to the lowest possible value of MISE of the estimated density (Eq. (5)) and given by,

$$MISE_Z(h,\lambda) = E\int \left\{\hat{f}_Z(z;h,\lambda) - f_Z(z)\right\}^2 dz \tag{6}$$

Assume that the first and second derivatives of the function $f_Z(z)$ exist, as well as that $K_1 = \int Z^2K(Z)\,dz$, $K_2 = \int K^2(Z)\,dz$ so

$$MISE_Z(h,\lambda) = AMISE_Z(h,\lambda) + O\left(h^4 + n^{-1}h^{-1}\right) \tag{7}$$

Where:

$$AMISE_Z(h,\lambda) = h^4\left(K_1^2/4\right)\int \psi'\left\{\psi^{-1}(u)\right\} f_U''(u;\lambda)^2 du + n^{-1}h^{-1}K_2\, E\,\psi'(z) \tag{8}$$

and the minimized window width for any value of λ is given by,

$$h^*_{\lambda,z} = \left[\frac{K_2\, E\,\psi'(z)}{K_1^2\int \psi'\left\{\psi^{-1}(u)\right\} f_U''(u;\lambda)^2 du}\right]^{1/5} .n^{-1/5} \tag{9}$$

and it contains less **AMISE**$_z(.,\lambda)$ for each constant value of λ that equals,

$$\underset{h>0}{Inf}\, AMISE_z(h,\lambda) = (5/4)\left(K_1K_2^2\right)^{2/5} J_z(\lambda)n^{-4/5} \tag{10}$$

Where,

$$J_z(\lambda) = \left[\left\{E\,\psi'(z)\right\}^4 \int \psi'\left\{\psi^{-1}(u)\right\} f_U''(u;\lambda)^2 du\right]^{1/5} \tag{11}$$

The last two equations (Eqs. (10) and (11)) represent a measure of the transformation's $\psi(z)$ influence on minimizing the error associated with estimating the function of the original data $\hat{f}_z(.;h,\lambda)$. Therefore, the optimal value of λ can be known as the one that minimizes: $\underset{h>0}{Inf}\, AMISE_z(h,\lambda)$.

By the same decision rules logic $AMISE_z(h,\lambda)$, derived from the density estimation of the transformed variable Z, the optimal asymptotical window width for each λ according to the original random variable is:

$$h^*_{\lambda,u} = \left[\frac{K_2}{K_1^2 J_u(\lambda)n}\right]^{1/5} \tag{12}$$

Asymptotically, the optimal choice of λ minimizes

$$\underset{h>0}{Inf}\, AMISE_u(h,\lambda) = (5/4)\left(K_1K_2^2\right)^{2/5} J_u(\lambda)n^{-4/5} \tag{13}$$

Where:

$$J_u(\lambda) = \left[\int f''_U(u;\lambda)^2 du\right]^{1/5} \tag{14}$$

In other words, it can be said that the minimization of $J_z(\lambda)$ and $J_u(\lambda)$ are the sufficient condition to prove the optimization of λ since Eq. (11) and Eq. (14) represents the variable parts of Eq. (10) and Eq. (13) respectively.

Finally, the relationship between $MISE_u(h,\lambda)$ and $MISE_z(h,\lambda)$ can be determined according to the equations:

$$MISE_z(h,\lambda) = E\int \left\{\hat{f}_u(u) - f_u(u)\right\}^2 \psi'\{\psi^{-1}(u)\}du \tag{15}$$

$$MISE_u(h,\lambda) = E\int \left\{\hat{f}_z(u) - f_z(u)\right\}^2 \psi^{-1})'\{\psi(z)\}dz. \tag{16}$$

Both error functions yield the same results, whether in terms of the original variable or of the transformed variable.

3. A new approach of transformations in KDS

Unlike BCT methodology, which assumes that the original data is of unknown distribution, PTs' in KDF estimation are used to shifted the random variables with a known distribution into symmetric shapes to obtain an efficient kernel density estimation. The statistical literature in nonparametric estimation suggested the use of MISE indicator as a decision rule for power parameter estimation for a number of distributions such as Lognormal [8, 18], gamma [8], Cauchy [9], Pareto [18] and heavy-tailed distributions [19, 20].

Now, similar to the BCT approach, the primary hypothesis of the new approach in this chapter is that the data do not have a definite distribution. We will use the power transformation to transform the data to a normal shape and use MISE as a decision rule to choose the optimal value of the power parameter. Later in the sections 4 and 5 we will use this approach in the functional nonparametric time series analysis.

Let us assume that we have the random variable Z with unknown distribution and $U = \psi(z)$ represents a PT model. Let, for Finney transformation (FT), suppose $U = z^{\lambda}$ follows the normal distribution with mean μ and variance σ^2. Therefore, according to the Eq. (3), the PDF of the original variable Z is given by $f_Z(z) = \psi'(z) f_U(\psi(z_i);\mu,\sigma^2)$.

In our proposed approach, the assumption of the normality of the transformed data when the original data is of unknown distribution provides uncomplicated options for estimating the power parameter so that Eq. (14) can be used as the simplest alternative to Eq. (11). In our assumption, we have,

$$f_U(u;\lambda) = \frac{1}{\sqrt{2\pi\sigma^2}} e^{\frac{-(u-\mu)^2}{2\sigma^2}}, U \in R \tag{17}$$

So, the square of the second derivative of Eq. (17) is,

$$f''(u;\lambda)^2 = \left(\sqrt{2\pi\sigma^2}\right)^{-2} exp\left(\frac{-(u-\mu)^2}{\sigma^2}\right)\left[\frac{1}{\sigma^4} + \frac{-2}{\sigma^2}\frac{(u-\mu)^2}{\sigma^4} + \frac{(u-\mu)^4}{\sigma^8}\right] \tag{18}$$

By inserting the integration factor, we get,

$$\int f''(u;\lambda)^2 du = \sigma^{-8}\left(\sqrt{2\pi\sigma^2}\right)^{-1}\left[\int \sigma^4\left(\sqrt{2\pi\sigma^2}\right)^{-1} e^{\frac{-(u-\mu)^2}{\sigma^2}} du - \int 2\sigma^2(u-\mu)^2\left(\sqrt{2\pi\sigma^2}\right)^{-1} e^{\frac{-(u-\mu)^2}{\sigma^2}} du + \int (u-\mu)^4\left(\sqrt{2\pi\sigma^2}\right)^{-1} e^{\frac{-(u-\mu)^2}{\sigma^2}} du\right] \tag{19}$$

Assume $\sigma^2 = 2\delta^2$, then the first term of Eq. (19),

$$\int \sigma^4 \frac{1}{\sqrt{2\pi 2\delta^2}} e^{\frac{-(u-\mu)^2}{2\delta^2}} du = \frac{\sigma^4}{\sqrt{2}} \int \frac{1}{\sqrt{2\pi\delta^2}} e^{\frac{-(u-\mu)^2}{2\delta^2}} du = \frac{\sigma^4}{\sqrt{2}} \tag{20}$$

and the second term of Eq. (19),

$$\begin{aligned}\int 2\sigma^2(u-\mu)^2\left(\sqrt{2\pi\sigma^2}\right)^{-1} e^{\frac{-(u-\mu)^2}{\sigma^2}} du &= \frac{2\sigma^2}{\sqrt{2}}\int (u-\mu)^2\left(\sqrt{2\pi\delta^2}\right)^{-1} e^{\frac{-(u-\mu)^2}{2\delta^2}} du \\ &= \frac{2\sigma^2}{\sqrt{2}} E(U-\mu)^2 = \frac{\sigma^4}{\sqrt{2}}\end{aligned} \tag{21}$$

and the third term of Eq. (19),

$$\begin{aligned}\int (u-\mu)^4\left(\sqrt{2\pi\sigma^2}\right)^{-1} e^{\frac{-(u-\mu)^2}{\sigma^2}} du &= \frac{1}{\sqrt{2}}\int (u-\mu)^4\left(\sqrt{2\pi\delta^2}\right)^{-1} e^{\frac{-(u-\mu)^2}{2\delta^2}} du \\ &= \frac{1}{\sqrt{2}} E(U-\mu)^4\end{aligned} \tag{22}$$

by using the central moments equation of the real-valued random variable U, $E(U-\mu)^n = E\sum_{j=0}^{n} C_j^n (-1)^{n-j}\, U^j\, \mu^{n-j}$ then,

$$E(U-\mu)^4 = E\left(U^4\right) - 4\,\mu\, E\left(U^3\right) + 6\,\mu^2 E\left(U^2\right) - 4\,\mu^3\, E(U) + \left(\mu^4\right) \tag{23}$$

Based on the moments equation, $\mu_k = E\left(U^k\right)$, get,

$$E(U-\mu)^4 = \mu_4 - 4\,\mu_3\,\mu + 6\,\mu_2\,\mu^2 - 4\,\mu\,\mu^3 + \mu^4 \tag{24}$$

Substitute the three parts defined by Eq. (20), Eq. (21) and (Eq. (22) into Eq. (19) get,

$$J_U(\lambda) = \left[\frac{1}{\sigma^8}\left(\sqrt{2\pi\sigma^2}\right)^{-1}\left(\frac{1}{\sqrt{2}}\left(\mu_4 - 4\mu_3\mu + 6\mu_2\mu^2 - 4\mu\mu^3 + \mu^4\right)\right]^{1/5} \tag{25}$$

Eq. (25) is the end of the derivation. The optimal power parameter value is the one that minimizes the value of $J_U(\lambda)$. In the practical application, the estimators of the maximum likelihood method were used for the moments about zero $\hat{\mu}^k = \sum_{i=1}^{n} u_i^k / n$ and the central moments $\hat{\mu}_k = \sum (u_i - \overline{u})^k / n$.

DOI: http://dx.doi.org/10.5772/intechopen.105832

4. Proposed application algorithm

For the univariate time series $\{Z_t, t \; R\}$, assume that the sample is divided into $(p+1))$ statistical samples of size $(n = N - s - p + 1)$ so that the time series data set can be defined as a functional data $\{(X_i, Y_i)\}_{i=1,..,n}$. The regression model,

$$Y = m(X) + \varepsilon \tag{26}$$

represents the relationship between the smooth functional data $m(X)$ and scalar response $Y_i = Z_{i+s}, i = p, \; , N - s$. The white noise ε is a sequence of independent identically distributed functions in such $E(\varepsilon/X) = 0$. $X_1, X_2, \dots, X_n$ are identically distributed as the functional random variable $X_i = \left(Z_{i-p+1}, \dots, Z_i\right)$. Assume $\mathbf{N} = \mathbf{n\tau}$ for some $n \in N$ and some $\tau > 0$ to get a statistical sample of curves $X_i = \{Z(t), (i-1)\tau < t \;\; i\tau\}$ of size $(n-1)$ and the response $Y_i = Z(i\tau + s), i = 1, \dots, n-1$ [5]. The kernel regression estimator evaluated at a given function $m(X)$ in Eq. (26) by:

$$\hat{m}(X) = \frac{\sum_{i=1}^{n} Y_i K\left(h^{-1} \, d\,(X, X_i)\right)}{\sum_{i=1}^{n} K\left(h^{-1} \, d\,(X, X_i)\right)} \tag{27}$$

Where K is a kernel function and, h (depending on n) are a positive real bandwidth and $d\,(X, X_i)$ denotes any semi-metric index of proximity between the observed curves based on the functional principal components [5, 6, 21]. Many authors have proposed a number of methods for measuring the proximity such as, the method of FPCA in which, $d\,(X, X_i)$ is measuring by the square root of the quantity $\int \left(X_i(t) - X_j(t)\right)^2 dt$ or the quantity $\int \left(X_i^{(2)}(t) - X_j^{(2)}(t)\right)^2 dt$ (for more details, see [4, 21–25]).

The application methodology includes estimating the smooth functional data $m(X)$ in the regression equation Eq. (26) according to the kernel estimator Eq. (27) after transforming the time series dataset. So, the following proposed application algorithm of the nonparametric estimation of transformed functional time series according to the proposed new approach for transforming the kernel density were as follows:

Step 1: Choosing the common range $\Lambda = \{-3, 3\}$ for the power parameter λ

Step 2: Calculate the value of $J_{\psi(Z)}(\lambda)$ according to Eq. (14).

Step 3: Transform the original response variable Z according the Finney [11] PT model, $\Psi(Z) = Z^\lambda$ when $\lambda \neq 0$ and BCT model Eq. (1) to get the explanatory functional matrices $\Psi_\lambda(X) = [\Psi_\lambda(z)]_{nx\tau}$ (for more about the matrices file organizing in the R program, see [5, 21]".

Step 4: Redefining the functional data of the regression model $\{(X_i, Y_i)\}_{i=1,..,n}$ so that the statistical sample of curves $X_i = \{Z(t), (i-1)\tau < t \;\; i\tau\}$ is defined as follows,

$$\Psi_\lambda(X_i) = \{\Psi_\lambda(Z(t)), (i-1)\tau < t \leq i\tau\} \tag{28}$$

and the response $Y_i = Z(i\tau + s), i = 1, \dots, n-1$ is defined as follows,

$$\Psi_\lambda(Y_i) = \Psi_\lambda(Z(i\tau + s)) \tag{29}$$

Step 5: Defining the Eq. (28) and Eq. (29) in which $\boldsymbol{\tau}$ equal the seasonal length.

Step 6: Estimate the explanatory function regression $\boldsymbol{\Psi}_\lambda(\boldsymbol{Y}_i) = \boldsymbol{m}(\boldsymbol{\Psi}_\lambda(\boldsymbol{X})) + \boldsymbol{\varepsilon}$, (where)

$$\hat{\boldsymbol{m}}(\boldsymbol{\Psi}_\lambda(\boldsymbol{X})) = \frac{\sum_{i=1}^{n} \boldsymbol{\Psi}_\lambda(\boldsymbol{Y}_i)\boldsymbol{K}\left(\boldsymbol{h}^{-1}\,\boldsymbol{d}\,(\boldsymbol{\Psi}_\lambda(\boldsymbol{X}), \boldsymbol{\Psi}_\lambda(\boldsymbol{X}_i))\right)}{\sum_{i=1}^{n} \boldsymbol{K}\left(\boldsymbol{h}^{-1}\,\boldsymbol{d}\,(\boldsymbol{\Psi}_\lambda(\boldsymbol{X}), \boldsymbol{\Psi}_\lambda(\boldsymbol{X}_i)\right)} \tag{30}$$

by using the Nadaraya–Watson regression estimator for functional data.

Step 7: Perform the steps 2 through 6 for all $\boldsymbol{\lambda} \in \boldsymbol{\Lambda}$.

Step 8: Choose the optimal value that corresponds to the lowest value of $\boldsymbol{J}_U(\boldsymbol{\lambda})$.

Step 9: Calculate the estimator of mean square errors of the last curve $\boldsymbol{MSE}(\boldsymbol{X}_n) = (\mathbf{1}/\boldsymbol{s})\sum_{j=1}^{s}(\hat{\boldsymbol{z}}_j - \boldsymbol{z}_j)^2$, where, $\hat{\boldsymbol{z}}_j$ and $\mathbf{z}_j$ are the j-th estimated and real values respectively in the last curve. $\hat{\boldsymbol{z}}_j$values denoted to, they are computed from the back transform of $\boldsymbol{\psi}(\boldsymbol{z}) = \boldsymbol{z}_i^{\lambda}$.

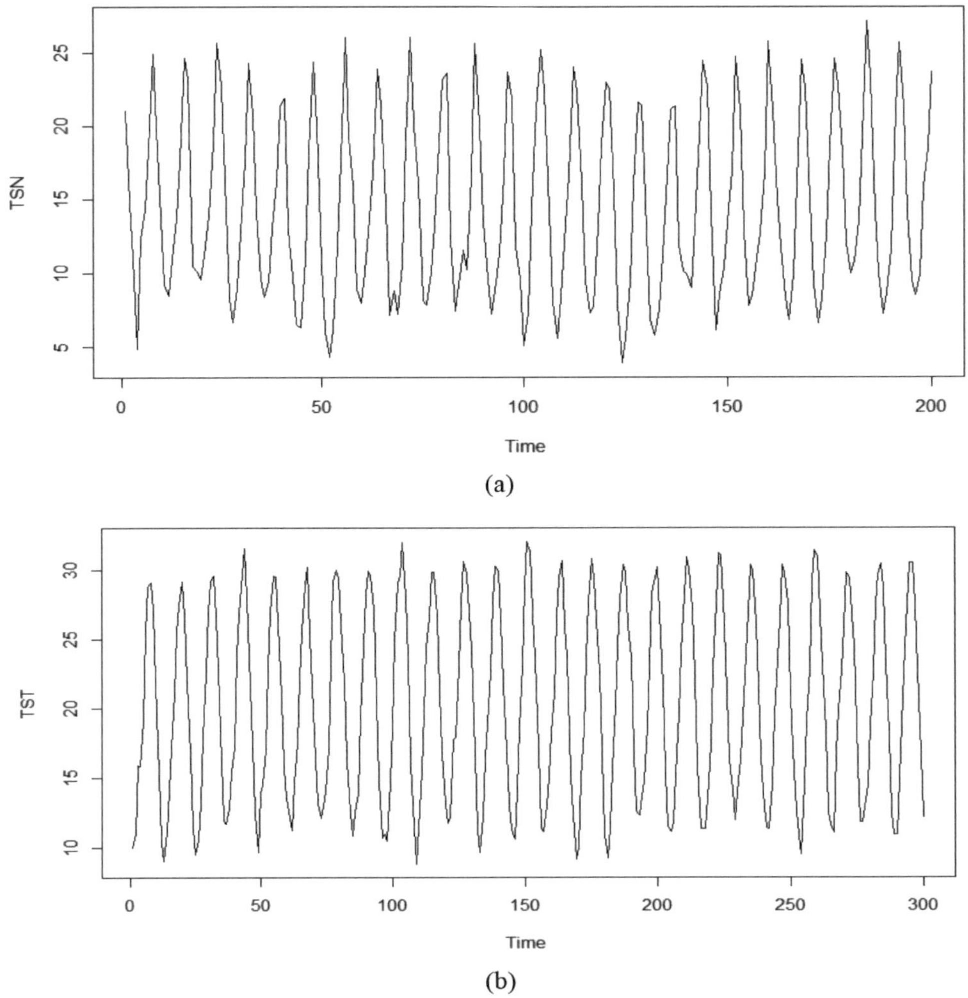

Figure 1.
Plots of the monthly temperature averages series: (a) TSN; (b) TST [21].

In all PT methodologies, the decision rule for choosing the optimal power parameter, always leads to what we might call the area of feasible solutions. For example, the argumentative question in BCT is: Does the optimal parameter that results from minimizing MLE method for the original response function achieve the normality of the transformed variable in practice? The authors believe that this problem is due to the nature of the data. In the proposed approach, the optimal power parameter that corresponds to the lowest $J_U(\lambda)$, we have the challenge of complexity in the feasible solutions area that we suppose to achieve: The transformed response normality in practical application that provides quality conditions for both functional and non-parametric analyzes approaches in nonstationary seasonal time series (For more see [26, 27] that point to other challenges related to the use of PT and the quality of the power parameter estimation).

5. Applications

The PT models indicated in the proposed application algorithm have been applied to two examples of nonstationary time series of monthly temperature averages [21]. The first has a size of 200 observations of Nineveh City in Iraq (TSN) for the period 1976 to 2000 (**Figure 1a**). The second has a size of 300 observations of Tunisia (TST) of the period 1991 to 2015 (**Figure 1b**). R software was used to analyze the data. The data is available at https://climateknowledgeportal.worldbank.org.

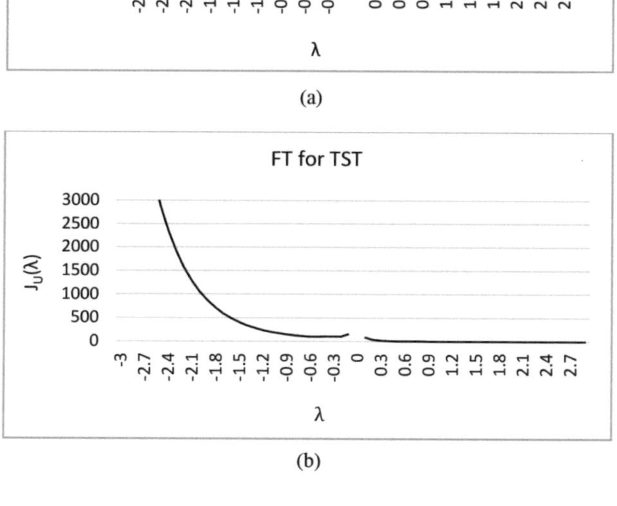

Figure 2.
The curves of the ordered pairs $\left(\hat{\lambda}_i, J_u(\lambda)\right)$ of the transformed responses of the two time series data sets using FT: (a) TSN. (b) TST.

Returning to the ideas of the of feasible solutions area, we must verify the results of choosing the optimal PT value according to the proposed density transformation approach and its contribution to achieving the analysis efficiency requirements: the concavity of $\boldsymbol{J_u}(\lambda)$, the normality of the transformed response, and the reduction of the prediction error in the functional nonparametric time series analysis.

Mathematically, $\boldsymbol{J_u}(\lambda)$ is a concave function, but a number of authors state the possibility that there is no mini-point or is not unique [8, 18]. This conclusion may depend on the success in choosing the appropriate PT model [20]. The plots in **Figures 2** and **3** show the curves of the ordered pairs $(\lambda_i, \boldsymbol{J_u}(\lambda))$ of Eq. (14).

In **Figure 2**, it can be seen that the curves of the two time series data sets using FT has a concavity point in the range $(-\mathbf{3}, \mathbf{0})$, while the $\boldsymbol{J_u}(\lambda)$ values tends to zero in which the curves fades towards the horizontal line in the range $(\mathbf{0}, \mathbf{3})$.

While when applying BCT, it becomes clear from **Figure 3** that there is no point of concavity in the curves of $\boldsymbol{J_u}(\lambda)$ as its value goes to zero whenever the value of λ goes to -3. Therefore, it is not possible to obtain an optimal value for λ.

As for the normality of the transformed data, **Table 1** shows for the two examples, that the response variable data in its original and transformed states are not normal. Both optimal values of λ corresponding to the minimum values of $\boldsymbol{J_u}(\lambda)$ did not shift the data to the normal shape. But on the other hand, the improvement in the forecastability of the two-time series was evident through the estimates of mean square errors of the last curve (**Table 2**).

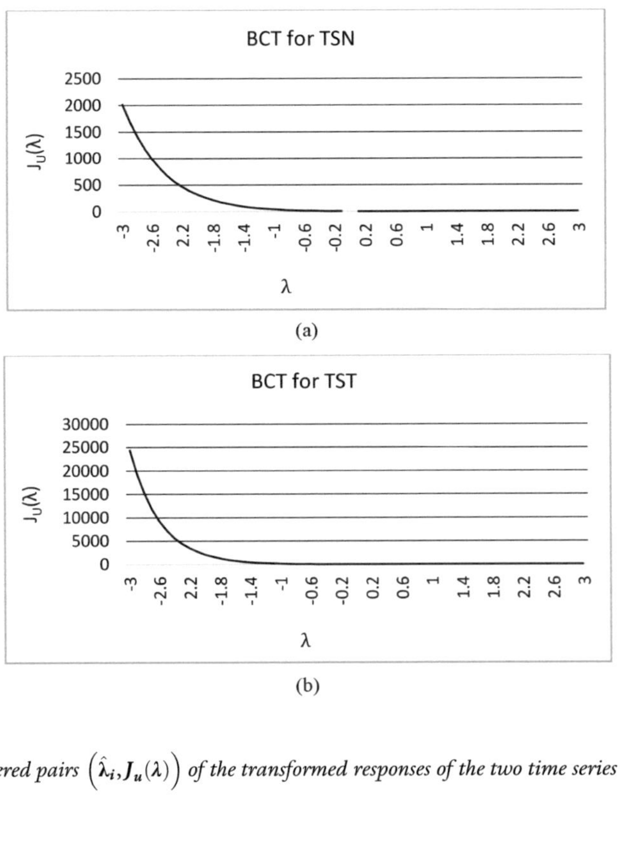

Figure 3.
The curves of the ordered pairs $\left(\hat{\lambda}_i, \boldsymbol{J_u}(\lambda)\right)$ *of the transformed responses of the two time series data sets using BCT: (a) TSN. (b) TST.*

Time Series	Responses	$\hat{\lambda}$	p-value	
			K-Smirnov	Sh.-Wilk
TSN	$Z(t)$	1.0	0.0002	8.5E–7
	$\Psi_{\lambda}(Z(t))$	−0.4	2.2E–16	2.8E–6
TST	$Z(t)$	1.0	2.0E–8	5.0E–11
	$\Psi_{\lambda}(Z(t))$	−0.6	2.2E–16	5.9E–10

Table 1.
The data normality tests of the original and transformed responses in the two examples using FT model.

Time Series	Responses	$\hat{\lambda}$	$MSE_Z(X_n)$
TSN	$Z(t)$	1.0	1.7616
	$\hat{\psi}^{-1}(Z(t))$	−0.4	0.9462
TST	$Z(t)$	1.0	0.4303
	$\hat{\psi}^{-1}(Z(t))$	−0.6	0.1994

Table 2.
The MSE estimates of the last curve $\mathbf{X_n}$ of the two-time series datasets.

6. Conclusions

In the analysis of parametric and non-parametric time series, like any statistical modeling process that requires the availability of certain conditions so that the results of statistical inference are reliable, which contributes to improving the forecastability.

Data is rarely ready for statistical analysis, which necessitates the use of power transformation to improve the required output. In this chapter, power transformation has been used with a new methodology to improve the outputs of the analysis with the following three directions: time series, nonparametric estimation and functional analysis. Therefore, the authors faced the challenge of choosing the optimal power parameter estimation method in accordance with the conditions of the feasible solutions area for the three directions. Using MISE as a criterion for choosing the power parameter in the proposed method did not achieve the normality of the data but it enhanced the forcasetibility of the time series.

By applying the FT and BCT models, the first was applicable and fulfilled the concavity condition for the transformation effect measurement function $J_u(\lambda)$, while the function curve was divergent at both ends of the power parameter range using the second model.

In the future, we recommend developing the proposed methodology using other transformation models and looking into the possibility of using it in other shapes of time series.

Author details

Haithem Taha Mohammed Ali[1,2] and Sameera Abdulsalam Othman[3]*

1 Department of Economic Sciences, University of Zakho, Kurdistan Region, Iraq

2 Department of Economic, Nawroz University, Kurdistan Region, Iraq

3 Department of Mathematics, College of Basic Education, University of Duhok, Kurdistan Region, Iraq

*Address all correspondence to: sameera.othman@uod.ac

References

[1] Germán Aneiros-Pérez G, Cao R, Vilar-Fernández JM. Functional methods for time series prediction: A nonparametric approach. Journal of Forecasting. 2010;**30**(4):377-392. DOI: 10.1002/for.1169

[2] Kidziński Ł. Functional time series. preprint arXiv:1502.07113. [stat.ME] 2015. https://doi.org/10.48550/arXiv.1502.07113

[3] Kannel PR, Lee S, Kanel SR, Khan SP. Chemometric application in classification and assessment of monitoring locations of an urban river system. Analytica Chimica Acta. 2007; **582**(2):390-399. DOI: 10.1016/j.aca.2006.09.006

[4] Dauxois J, Pousse A, Romain Y. Asymptotic theory for the principal component analysis of a vector random function: Some applications to statistical inference. Journal of Multivariate Analysis. 1982;**12**(1):136-154

[5] Ferraty F, Vieu P. Nonparametric models for functional data, with application in regression, time series prediction and curve discrimination. Nonparametric Statistics. 2004;**16**(1–2): 111-125

[6] Shang H, Xu R. Functional time series forecasting of extreme values. Communication Statistics. 2021;7(2): 182-199

[7] Maadooliat M, Huang JZ, Hu J. Integrating data transformation in principal components analysis. Journal of Computing Graphical Statistics. 2015; **24**(1):84-103. DOI: 10.1080/10618600.2014.891461

[8] Wand MP, Marron JS, Ruppert D. Transformations in density estimation. Journal of the American Statistical Association. 1991;**86**(414):343-353

[9] Ruppert D, Wand MP. Correcting for kurtosis in density estimation. Australian and New Zealand Journal of statistics. 1992;**34**(1):19-29

[10] Chavez-Demoulin V, Davison AC. Modelling time series extremes. REVSTAT-Statistical Journal. 2012; **10**(1):109-133

[11] Finney DJ. Statistical Method in Biological Assay, Charles Griffin. 1st ed. London: Charles Griffin &; Co. Ltd; 1952

[12] Box GEP, Cox DR. An Analysis of Transformations. Journal of the Royal Statistical Society. Series B (Methodological). 1964;**26**(2):211-252

[13] Tukey JW. On the comparative anatomy of transformations. The Annals of Mathematical Statistics. 1957;**28**: 602-632. DOI: 10.1214/aoms/1177706875

[14] Yang L, Marron JS. Iterated transformation–kernel density estimation. Journal of the American Statistical Association. 1999;**94**(446):580-589

[15] Bean A, Xinyi X, MacEachern S. Transformations and Bayesian density estimation. Electronic Journal of Statistics. 2016;**10**(2):3355-3373

[16] Pitt D, Guillen M, Bolancé C. Estimation of parametric and nonparametric models for univariate claim severity distributions: An approach using R. Journal of Financial Education. 2011;**42**(1–2):154-175

[17] Sakthivel KM, Rajitha CS. Kernel density estimation for claim size distributions using shifted power

transformation. International Journal of Science and Research. 2013;**6**(14): 2025-2028

[18] Bolance C, Guillen M, Perch Nielsen J. Kernel density estimation of ac tuarial loss functions. Insurance Mathe matics and Economics. 2013;**32**(1):19-36

[19] Koekemoer G, Swanepoel JWH. Transformation Kernel density estimation with applications. Journal of Computational and Graphical Statistics. 2008;**17**(3):750-769

[20] Bean A. Transformations and Bayesian Estimation of Skewed and Heavy-Tailed Densities. Ohio State University; Ohio LINK Electronic Theses and Dissertation Center 2017. 2017. http://rave.ohiolink.edu/etdc/view?acc_num=osu1503015935192212

[21] Othman SA, Ali HTM. Improvement of the nonparametric estimation of functional stationary time series using Yeo-Johnson transformation with application to temperature curves. Advances in Mathematical Physics. 2021; **2021**. DOI: 10.1155/2021/6676400

[22] Castro PE, Lawton WH, Sylvestre EA. Principal modes of variation for processes with continuous sample curves. Technometrics. 1986; **28**(4):329-337

[23] Ferraty F, Vieu P. Curves discrimination: A nonparametric functional approach. Computational Statistics & Data Analysis. 2003; **44**(1–2):161-173

[24] Ferraty F, Vieu P. Nonparametric Functional Data Analysis: Theory and Practice. Springer Science & Business Media; 2006

[25] Febrero-Bande M, de la Fuente MO. Statistical computing in functional data analysis: The R package fda. usc. Journal of statistical Software. 2012;**51**(1):1-28

[26] Atkinson AB, Riani M, Corbellini A. The Box–Cox transformation: Review and extensions. Statistical Science. 2021

[27] Soleymani S. Exact Box-Cox Analysis. The University of Western Ontario; Electronic Thesis and Dissertation Repository, 2018. https://ir.lib.uwo.ca/etd/5308/

Section 6

Signal Detection and Monitoring

Chapter 6

Change Detection by Monitoring Residuals from Time Series Models

Tom Burr and Kim Kaufeld

Abstract

Change detection in time series can be approached by fitting a model to the no-change, ordinary background data and then monitoring time series of residuals, where a residual is defined as residual = data – fit. In many applications, models that fit time series data lead to residuals that exhibit no patterns unless the signal of interest is present. Therefore, an effective signal or change detection approach is to first fit a time series model to the background data without any signal and then monitor the time series of residuals for evidence of the signal. This chapter briefly reviews a few time series modeling options and then focuses on statistical tests for monitoring residuals, including Page's cumulative sum (cusum, a type of scan statistic), the ordinary cumulative sum (cumsum), the matched filter (a version of the Neyman-Pearson test statistic), and pattern tests, such as those used in quality control. Simulation and analytical approximation methods are recommended for studying test behavior, as illustrated in three examples.

Keywords: time series models, residuals, scan statistics, cusum, matched filter

1. Introduction

This chapter's focus is on change detection by monitoring residuals arising from fitted models to time series data. The residual at time t is defined as $r_t = x_t - \hat{x}_t$, where $\hat{x}_t$ is the predicted (estimated) value of the data x_t. Large residuals or patterns in residuals could indicate that some type of change has occurred compared to usual behavior during the analysis period that was used to fit the model. For example, the number of positive test results for a disease could show a sharp rise or decline compared to the recent past, perhaps indicating that a signal of interest is present, such as a more infectious strain emerging. As another example, assembly line productions monitor product quality, such as the diameter of a machined part, which can drift due to measurement effects and/or machining effects. Diameter drifting can lead to detectable residual patterns, where the residual is the measured diameter—target diameter. Prior to diameter drifting, the time series should vary randomly around a mean value that is close to the target mean diameter. Therefore, time series fitting of the "in control" process data simply requires estimation of the mean and standard deviation of the measured diameter of each part. Other time series fitting options are less simple.

There are many types of time series, such as series of the unit or system failure times in reliability data, series of new disease cases or deaths, series of measured product quality, such as geometric dimensions in machined parts or salt content in bags of chips. This chapter does not consider predicting the next failure time, the time to the next spike in disease counts, or the time for the machined part mean dimension to shift. Instead, the chapter focuses on monitoring for possible changes in the mean-time to failure, changes in the distribution of disease counts, or machined part dimensions. There are many applications in which a training period for model fitting is assumed to define normal behavior, and then the testing period monitors for various types of changes from the normal behavior, such as a shift to a different mean value.

Figure 1 is a time series of $n = 50$ values that have mean 0 and standard deviation 1 (independently and identically distributed normal random values, denoted iid N(0,1) in the figure caption) that exhibit no change in (a), a mean shift of 2 units on period 25 in (b) a mean shift of 2 units at time indices 26–30 in (c), and a mean shift at time indices 26–50 in (d). The human brain/eye is reasonably effective at spotting such changes but is vulnerable to being fooled by spurious patterns. Statistical methods, some very simple and some less simple have been developed to detect changes of interest, as this chapter explains.

Let $x_1, x_2, \ldots, x_n$ denote a time series, which is a sequence of values at times $1, 2, \ldots, n$. **Figure 2** plots an example simulated time series with $n = 50$.

An effective time series model leads to residuals that are approximately independently and identically distributed (iid) [1, 2]. The residual at time t is defined

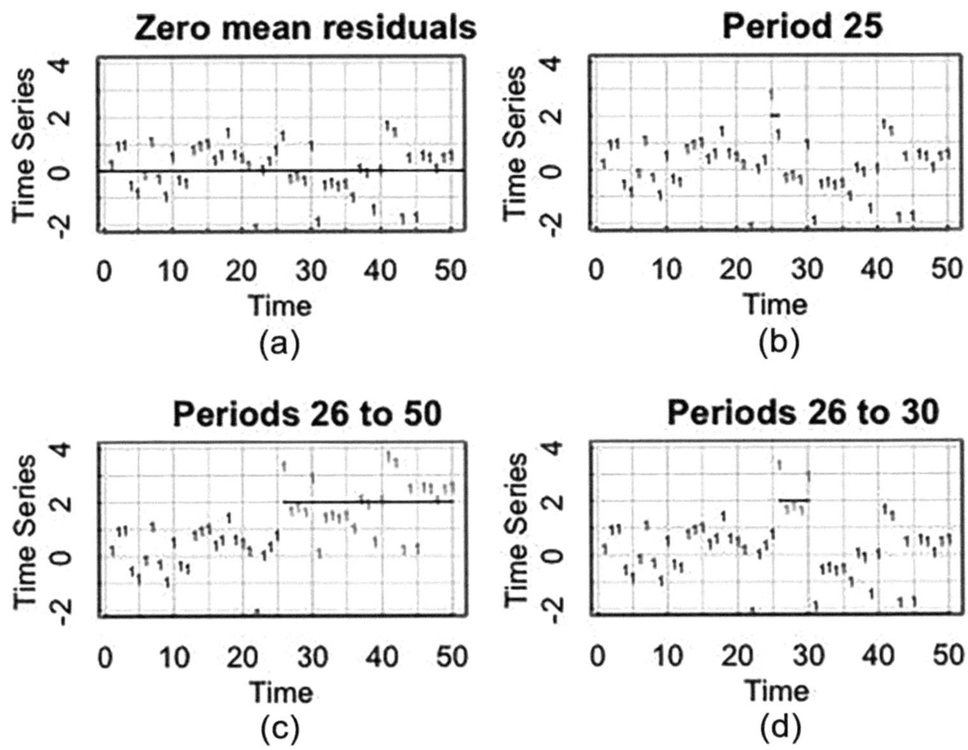

Figure 1.
Time series, iid N(0,1).

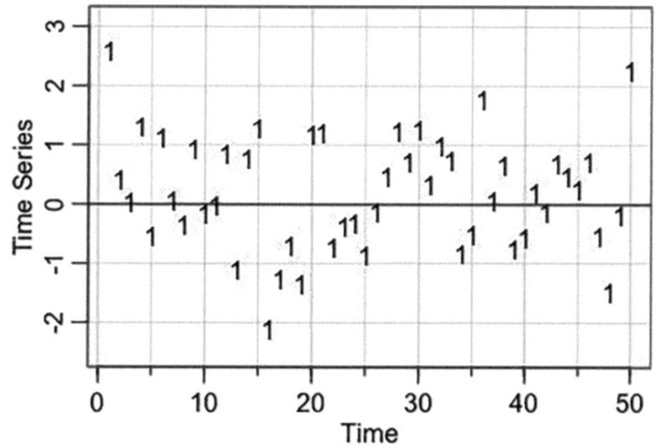

Figure 2.
Time series, MA(1) generated from iid N(0,1).

as $r_t = x_t - \hat{x}_t$, where $\hat{x}_t$ is the predicted (estimated) value of x_t. For example, $f()$ could be linear in the x's, resulting in the well-known auto-regressive (AR) model, $f() = \rho_1 x_{t-1} + \rho_2 x_{t-2} + \dots + \rho_{l_A} x_{t-l_A}$ or $f()$ might be linear in both the x's and an underlying iid noise sequence $e_1, e_2, \dots, e_n$, resulting in the auto-regressive moving average (ARMA) model $f() = \rho_1 x_{t-1} + \rho_2 x_{t-2} + \dots + \rho_{l_A} x_{t-l_A} + \theta_1 x_{t-1} + \theta_2 x_{t-2} + \dots + \theta_{l_M} x_{t-l_M}$ where l_A is the AR lag, and l_M is the moving average (MA) lag [1, 2]. The lag-one MA, AR, and ARMA models are given in (Eqs. (1)–(3)).

$$x_t = \theta_1 e_{t-1} + e_t \tag{1}$$

$$x_t = \rho_1 x_{t-1} + e_t \tag{2}$$

$$x_t = \rho_1 x_{t-1} + \theta_1 e_{t-1} + e_t \tag{3}$$

Conditions on the magnitudes of θ and ρ ensure stationarity (constant mean and variance over time). Often, time series can be transformed to stationarity by taking first differences, as is commonly done in stock market price series [1, 2]. Of course, $f()$ might not be linear in prior x values or e values and in general could be an arbitrarily complicated function, $f(x_{t-1}, \dots, x_{t-l_A}, e_{t-1}, \dots, e_{t-l_M})$.

Figure 2 is simulated data from a lag one MA model, $x_t = \theta_1 e_{t-1} + e_t$. **Figure 3** is the cumulative sum (cumsum, $C_t = \sum_{i=1}^{t} x_i$) and Page's cusum S_t with parameter k (see [3] and examples 2 and 3) defined as.

$$S_t = \max(0, S_{t-1} + x_i - k \tag{4}$$

for the data plotted in **Figure 2** [3]. Because of the reset-to-0 feature of S_t if the sum goes negative and because of the parameter k, Page's S_t does not have the large drift behavior that the cumulative sum does. **Figure 4** is the estimated underlying residual sequence and the actual underlying simulated residual sequence. **Figure 5** is the same as **Figure 4** but plots the difference between the estimated and true residuals. All plots and analyses are performed in R [4]. Statistical tests for changes in the background due to signal are applied to estimated residuals, so alarm thresholds and signal detection probabilities should be estimated using estimated residuals obtained via simulation.

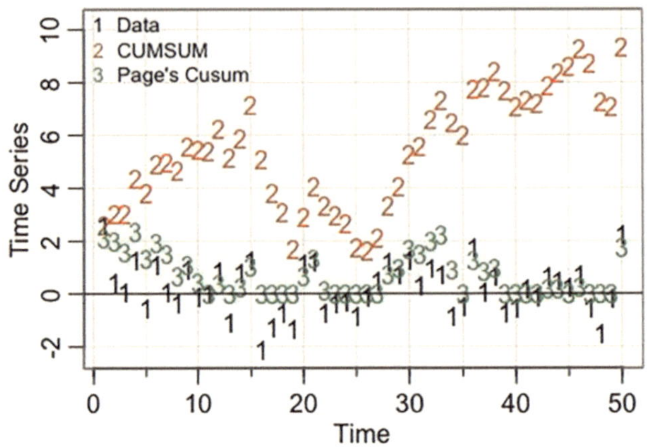

Figure 3.
IID N(0,1) time series, cumulative sum, and Page's Cusum.

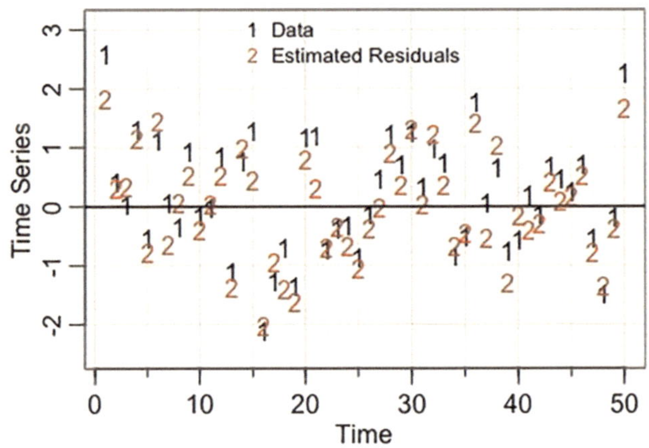

Figure 4.
Estimated residuals for MA(1) from N(0,1) data.

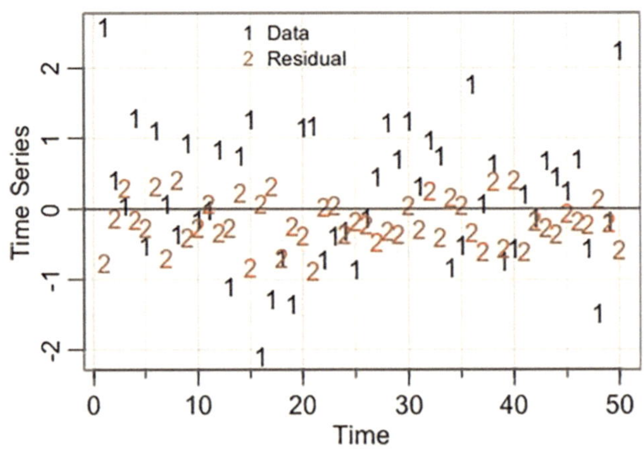

Figure 5.
*Same as **Figure 4**, but plotting the difference, residual-estimated residual.*

This chapter describes three change-detection examples. Example 1 monitors for patterns of large residuals, such as a consecutive string of three residuals exceeding a threshold. Example 2 monitors for excessive numbers of tweets in any of the 65 Florida counties. Example 3 monitors for nuclear material loss. Portions of Examples 1 and 3 have been published. Example 2 is entirely new.

2. Example 1: monitoring for patterns of large residuals

The Stein-Chen (SC) method approximates the probability density function (pdf) that assigns probabilities to the number of times that a pattern such as $I_t, I_{t+1}, I_{t+2} = \{1\ 0\ 1\}$ occurs, starting at position t in a binary time series of length n. In example 1, the original time series that is converted to binary is assumed to consist of a sequence of independent iid residuals that result from fitting any type of time series model. Recently the SC method was shown to provide an accurate Poisson-based approximation and corresponding total variation distance bounds in a time series context [5]. The binary values.

$I_1, I_2, I_3, \ldots, I_n$ are assumed to be independent and identically distributed with constant probability $p = P(I_i = 1)$. The probability p is the probability that the original time series X exceeds a threshold, and the I notation denotes an indicator or binary variable. As an aside, the SC method can also be applied if p is not constant over time, but the independence assumption is difficult to avoid [5–10]. Any type of time series model [1, 2] can be fit, and then the resulting residuals become the original series that is thresholded to convert to binary; therefore, the application is quite general. **Figure 6** is the estimated residuals from **Figure 4** but thresholded at 1.4 (values of 1.4 or larger are set to 1; values less than 1.4 are set to 0) to convert to binary.

Note that if $\{1\ 0\ 1\}$ is known to not occur, for example, starting at position $t = 1$, then this information impacts the probability that $\{1\ 0\ 1\}$ occurs starting at position $t = 2$ or $t = 3$, because the trials to obtain $\{1\ 0\ 1\}$are overlapping and thus not independent, so the Poisson distribution assumptions are not met. Nevertheless, Ref. [5] showed that Poisson-based approximation (that is strictly correct only for independent trials) can be remarkably accurate, and the SC method provides a bound on the total variation distance between the true and approximate pdf.

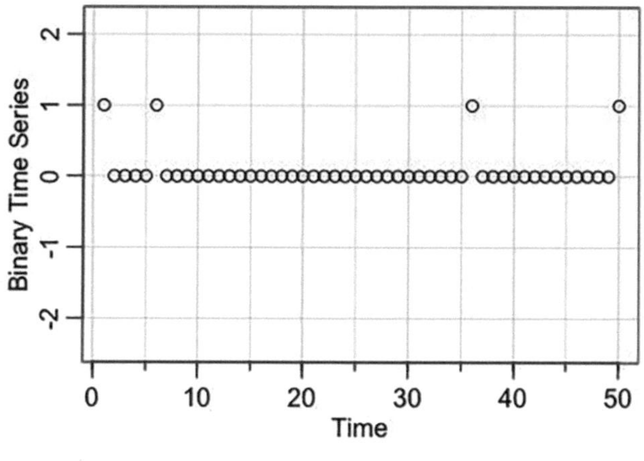

Figure 6.
Binary version of the estimated residuals in ***Figure 3****.*

Consider scanning for {1 x 1} with x = 0 or 1, with $p = P(I_i = 1)$ being quite small, such as 0.10 or less in a residual series of length n = 10. Then the probability of the pattern {1 x 1} is $p_p = p^2$, and there are $n - 2 = 8$ possible starting locations for the pattern in N = 10 trials. Because there are only 2^{10} = 1024 possible patterns of 0s and 1s, all 1024 patterns could be listed, and the probabilities assigned to each set of 10 binary values that include {1 × 1} at least once could be summed to provide an exact calculation. For larger values of n, this exact calculation is unwieldy, so an approximate method is desired, provided the approximation is highly accurate with provable error bounds.

Start at index i = 1 and check whether {1 × 1} occurs in positions {1 2 3}, then start at index i = 2 and check whether {1 × 1} occurs starting at index 2 in positions {2 3 4}, then start at index 3, etc. Note, for example, that if {1 × 1} occurs starting at position i = 1, then the probability that {1 × 1} also occurs starting at index 3 is p. Clearly, there is a small neighborhood of dependence around each starting index, as just illustrated. This neighborhood of dependence violates the assumptions for a Poisson distribution (as a limit distribution for a sequence of N Bernoulli trials, each with a small probability of success), but Ref. [6] shows that provided the dependence neighborhood is modest, the Poisson distribution can still provide an excellent approximation to the pdf defined on the number of times {1 × 1} occurs in a series of length N.

2.1 Stein-Chen method

The Poisson pdf with mean parameter $\lambda = (N-2)p_p$ provides an approximation Y to the true pdf W for the number of times {1 × 1} occurs in a series of length n [5, 6]. The value $(n-2)$ is used instead of n because the length 3 pattern could only be found starting at index 1, 2, … , $n-2$.

The quality of the Poisson(λ) approximation can be measured by computing b_1 and b_2, where $b_1 = \sum_{i=1}^{n}\sum_{j\in N_i} p_i p_j$, with $N_i = \{i-2, i-1, i, i+1, i+2\}$ being the dependent neighborhood N_i of index i, and $b_2 = \sum_{i=1}^{n}\sum_{j,j'\in N_i} E\{I_j I_{j'}\}$, where $j \neq j'$. The term b_3 is equal to 0 in Theorem 2 of Ref. [6] by the construction of N_i in this example. Then, the total variation distance (TVD) satisfies.

$$d_{TVD}(Y,W) \leq 4(b_1+b_2) = 4\left(n-2, 9p_p^2 + 3p_p p\right) \tag{5}$$

The TVD is a general distance measure between two pdfs. The TVD is defined here as the maximum absolute difference between the probability assigned by Y and the probability assigned by W to any specified subset of possible integer values. In the current scanning context, the most important subset of possible values to consider is the single value {0}, which would imply that the pattern {1 × 1} never occurred (occurred 0 times) in the $n-2$ overlapping trials. Then, the SC method, in this context, uses the Poisson approximation to assign a value to P {0} and this method ensures that the Poisson approximation to P{0} is quite accurate, as shown by the numerical example below.

According to the Poisson approximation, $P(\{1\,x\,1\}$ never occurs$) = e^{-\lambda}$. For example, using $n = 1000$ and $p = 0.01$, $\lambda = 998p_p = 0.0998$, then $e^{-\lambda} = 0.905$ is the approximate probability that the pattern never occurs, with an SC-based bound of $4(n-2)\left(9p_p^2 + 3p_p p\right) = 0.0123$. Therefore, the maximum difference between the true probability defined by the Y random variable and the approximate probability assigned to any subset of the possible number of occurrences of $\{1 \times 1\}$ defined by the approximating W (Poisson random variable) is 0.01236. So, for example, if the probability that ($\{1 \times 1\}$ never occurs) $= e^{-\lambda} = 0.905$, then the true probability of 0 occurrences of the pattern is between 0.89 and 0.92. Section 2.3 uses simulation to confirm the quality of the SC approximation in Eq. (5) in this context.

Simulation can be used to closely approximate the true probabilities, but only for moderate values of n. For example, in 10^6 repeated sets of $n = 1000$ Bernoulli trials with $p = 0.01$, then $\lambda = 998p_p = 0.0998$, and $e^{-\lambda} = 0.905$, the simulation-based P(0 occurrences of $\{1 x 1\}$) = 0.907. The Poisson-based approximation gives 0.905 with a SC-based TVD bound from Eq. (4) of 0.0123.

2.2 Summary

The SC method was described to approximate the pdf for the number of occurrences of an example pattern in an independent binary time series. In scanning for whether a pattern, such as $\{1 \times 1\}$, occurs starting at index i, there are overlapping tries to achieve the pattern, resulting in many non-independent trials consisting of the values in three successive indices. As the time series length increases and the probability $p = P(I_i = 1)$ decreases, the SC method shows that the Poisson approximation is excellent, with a small total variation distance bound.

The SC bound does not seem to be commonly used; however, related references are available [5–10]. For example, Ref. [7] applies the SC method to calculate coincidence probabilities. References [8, 9] apply the SC method in different time series contexts than considered here. To simplify the calculation of bivariate Poisson moments, Ref. [10] applies the SC identity $Xf(X) = \mu E(f(X+1))$, where X is a Poisson(μ) random variable, E denotes expected value, and $f()$ is any bounded function defined on the nonnegative integers. The SC identity was used to develop the SC approximation method used in Ref. [5] and the example above. Reference [5] showed that the SC bound defends the use of the Poisson approximation in real applications (as opposed to unwieldy combinatorial calculations for long time series), and provides a small bound on the approximation error. Simulation and/or analytical approximation are needed to estimate $p = P(I_i = 1)$ to apply the Poisson approximation and associated SC bound.

3. Example 2: monitoring for excessive numbers of tweets in 65 Florida counties

3.1 Introduction

Example 2 analyzes two types of daily tweet counts. The first type of counts is available from March 8, 2010, to December 31, 2015, in each of 65 Florida counties. In the available data from 65 counties (instead of the full 67), Lafayette is merged with Madison, and Liberty is merged with Gadsden; see Appendix 1. Such merging changes the spatial resolution available to detect spatial–temporal outbreaks for the four relevant counties. Reporting counts at the county (or merged county) level also changes the available spatial resolution, compared, for example, to geo-tagged counts. The second type of counts is time-tagged (to the nearest second) and geo-tagged (latitude and longitude) tweets, not aggregated to county-level counts.

One approach to monitor any numerically-valued time series has two main steps—(1) use training data to fit a model or models to the daily counts by county; and (2) monitor the corresponding residuals during training and testing to detect departures from the fitted model(s). For (1), effects such as day-of-week or time-of-year effects with multiple seasonal trends could be present, so the model building should be comprehensive, including assessment of simple exponentially weighted moving average (EWMA) models as well as models to fit trends and/or seasonal effects. For (2), the anomalies of interest could arise in neighboring counties (such as a reaction to a severe weather event such as a hurricane), and so could cluster in time and/or space.

3.2 Exploratory data analysis

Figure 7 is daily tweet counts from March 8 to December 31, 2010 (236 days) for (a) Brevard, (b) Broward, (c) Duval, and (d) Flagler counties (4 of 65 counties). This report addresses what types of models might fit these data, and options for monitoring residuals from the fitted model(s).

Question 1: Is the distribution of daily counts stable or stationary? Stationary means that, for example, the mean and variance of the counts is constant over time [1, 2]. Informally, it appears that spiking occurs in Broward county and that a mean shift occurs in Duval county. If the counts are not stationary, sometimes, for example, the first differences in counts are at least approximately stationary, perhaps with occasional spikes (**Figure 8**) [1, 2].

Question 2: Is the background ("training") data such as the 2010 counts adequately fit by a Poisson distribution [11–13]? **Figure 9** plots the variance/mean ratio (which

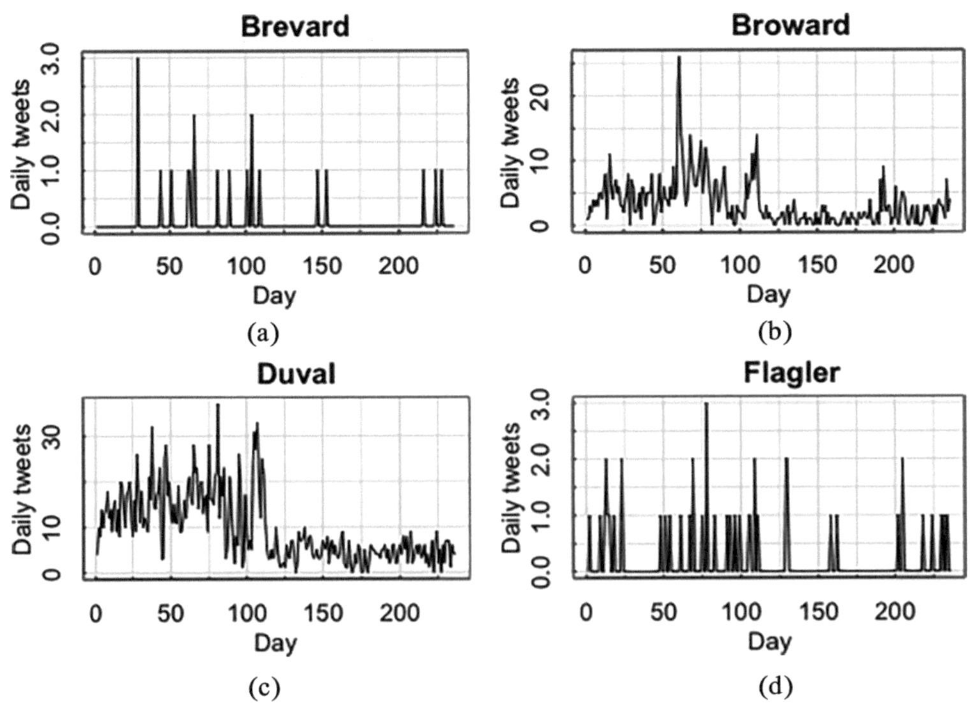

Figure 7.
The daily tweet counts for four of 65 Florida counties from march 8 to December 31, 2010.

should be approximately 1 for Poisson data) for each of the 65 counties. Hillsborough, Miami-Dade, and Orange counties appear to be outliers. In view of **Figure 9**, question 2 is academic here, because it is not expected that a simple constant-mean Poisson model will be adequate [11–13]. **Figure 9** suggests non-Poisson behavior.

Figure 10 plots the daily number of tweets for Broward county and simulated counts assuming a constant mean (equal to the mean of the Broward county counts) Poisson model. **Figure 10** also suggests non-Poisson behavior.

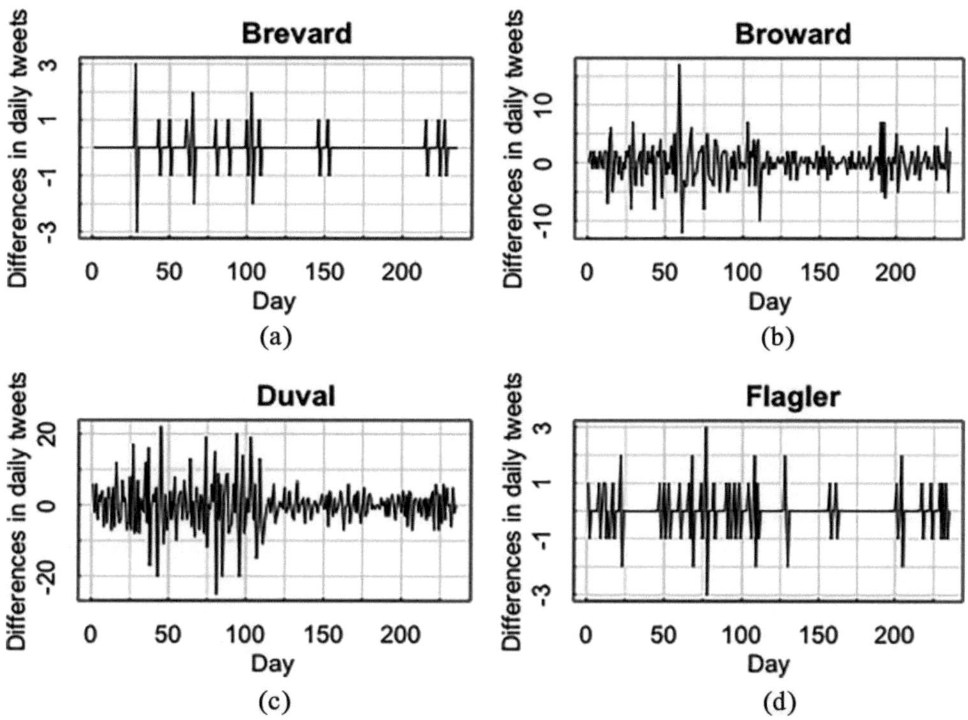

Figure 8.
The first differences in daily tweet count for four of 65 Florida counties from March 8 to December 31, 2010.

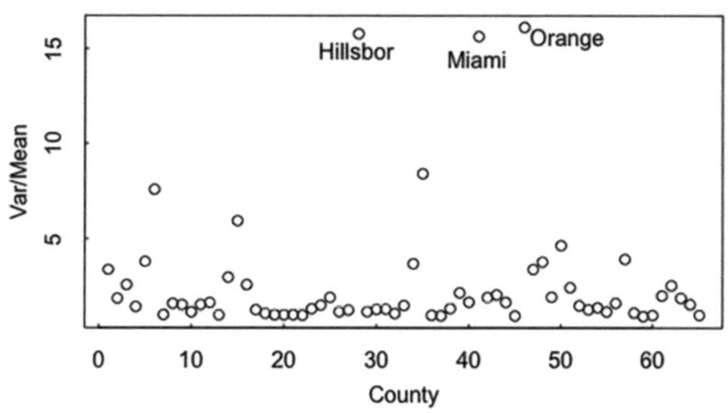

Figure 9.
The variance/mean for each of 65 Florida counties in 2010. The three largest variance/mean ratios are Hillsborough, Miami-Dade, and Orange counties.

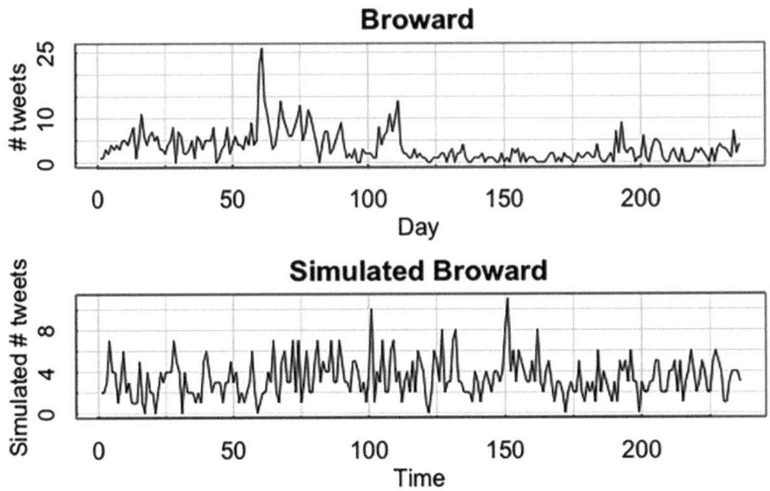

Figure 10.
The daily number of tweets in Broward county from March 8 to December 31, 2010, real (top), and simulated Poisson (bottom).

In **Figure 10**, the largest simulated count is 11 (the mean count is 3.4) but 7 of the 236 real counts exceed 11, and 2 of those exceed 20. More formally, the 99th percentile of the variance/mean ratio when sampling 236 observations from Poisson (3.4) is 1.23, but $\frac{s^2}{\overline{x}} = 3.4$ (**Figure 8**), where $\overline{x} = \frac{\sum_{i=1}^{n} x_i}{n}$ is the sample mean and $s^2 = \frac{\sum_{i=1}^{236} (x_i - \overline{x})^2}{235}$ is the sample variance. All analyses are performed using R [4]. Together, **Figures 9** and **10** strongly suggest that a constant mean Poisson model is not adequate. Possibly, residuals around a fitted model could display approximate Poisson behavior, or perhaps another model such as the negative binomial for which variance/mean > 1 would be more appropriate [11].

Figure 11 is the total counts overall 65 Florida counties for each year in 2010–2015. **Figure 12** is the autocorrelation function $ACF_i = \rho_i = \frac{\sum_{j=1}^{n} (x_i - \overline{x})((x_{j-i} - \overline{x})}{\sqrt{\sum_{j=1}^{n} (x_i - \overline{x})^2}}$ (with a minor adjustment to avoid negative indices [1]) using indices 1–117, prior to the large drop at index 118 for the total Florida counts in 2015. The ACF is useful for selecting possible models, such as those included in the class of ARMA models [1]. Modern machine-learning (ML) methods can also be evaluated, typically using AR modeling [1, 2, 12, 13]. An example of linear AR model is the lag one model $x_t = \mu + \rho x_{t-1} + e_t$, where μ is the long-term mean and the condition $|a| < 1$ ensures that the $\{x_t\}$ series is stationary. The noise term is denoted e_t. An example of a linear MA model is in Eq. (1); fitting an MA model requires estimating the noise sequence e_t [1, 2]. **Figure 13** is (a) simulated MA(1) data and (b) estimated residuals versus true residuals as was also shown in **Figures 4** and 5.

3.3 Model selection and fitting

Exploratory data analysis (EDA) indicates no significant correlations in the residuals from the ARIMA(0,1,1) fit, so a simple MA(1) model could be competitive. The

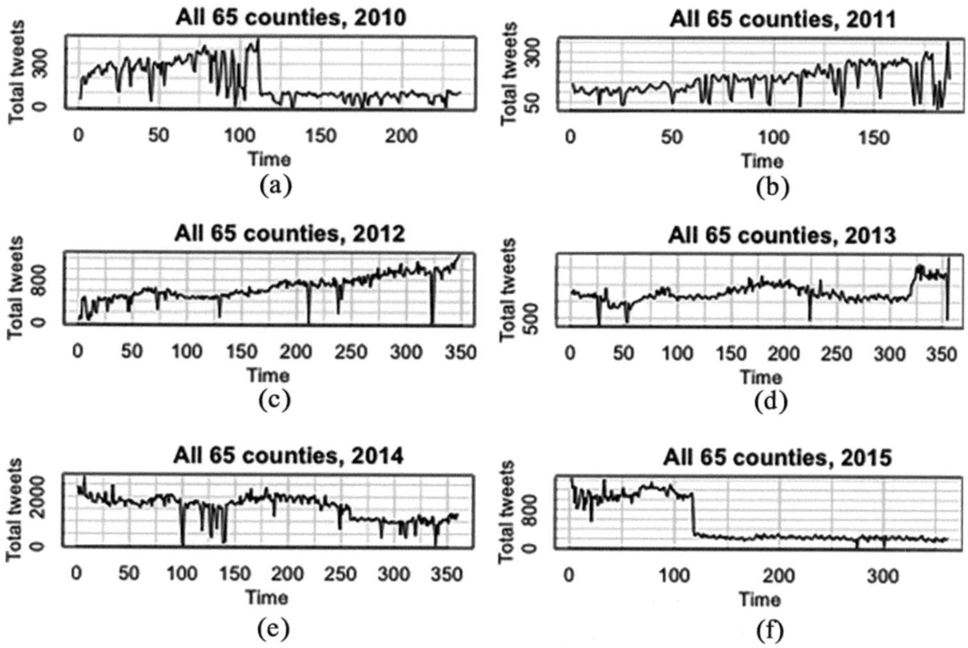

Figure 11.
The total counts overall 65 Florida counties for each year in 2010–2020,105.

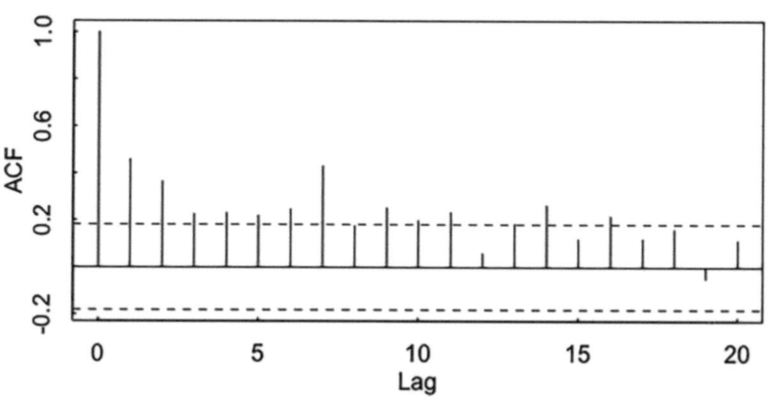

Figure 12.
The ACF for total Florida counts in 2015 over indices 1–117 before the large drop.

exponentially weighted moving average (EWMA) fit leads to the same fits as an MA (1) fit to the lag-one differences, which can be easily seen [1] as follows:

$$\hat{x}_t = (1-\lambda)\hat{x}_{t-1} + \lambda x_{t-1} = x_{t-1} + (1-\lambda)(x_{t-1} - \hat{x}_{t-1}) \tag{6}$$

$$x_t = (1-\lambda)e_{t-1} + e_t \tag{7}$$

$$\hat{x}_t = x_{t-1} - (1-\lambda)e_{t-1} \tag{8}$$

The forecast $\hat{x}_t$ in Eq. (6) from EWMA is the same as the forecast $\hat{x}_t$ in Eq. (7) from ARMA(0,1,1) which is an MA(1) after differencing [1, 2]. It should be pointed out that Poisson EWMA control charts as described in Ref. [11] are designed to monitor

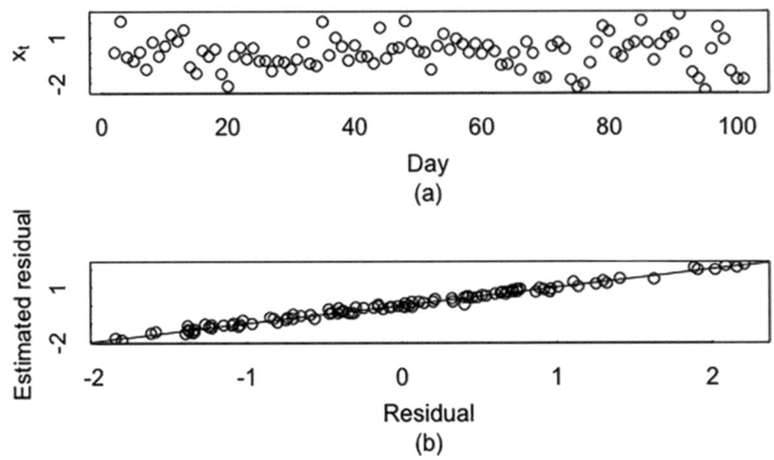

Figure 13.
(a) Simulated MA(1) data and (b) estimated residuals versus true residuals.

whether the Poisson mean μ_t is constant over time. In that case, $\hat{\mu}_t = (1-\lambda)\hat{\mu}_{t-1} + \lambda x_{t-1}$ is monitored rather than monitoring residuals from a local fit to the time-varying mean as in the twitter count monitoring.

The EWMA is among the simplest and most effective methods to forecast a time series. However, anomalies that persist for more than one day will impact the EWMA forecast in a manner that leads to reduced detection probability (DP) to detect the anomaly. For example, suppose an anomaly persists for 5 days. The EWMA forecast will be quite effective for day 1 of the anomaly, but the day 2 EWMA forecast will tend to increase due to the elevated expected count on day 1; this increased forecast leads to a reduced residual, thus reducing the DP.

Smooth fits using wavelets [13], EWMA [1, 2], and iterative bias reduction (IBR) [14–16] have been compared on this data. Edge effects are present in the wavelet and EWMA smoothers at the beginning of the data. The RMSE for wavelets, EWMA, and IBR are 4.05, 3.33, and 3.58, respectively, so EWMA is the simplest and has the smallest RMSE in this example. In data regions that have peaks, EWMA also performs acceptably well. The smoothing parameter λ in Eq. (6), $\hat{x}_t = (1-\lambda)\hat{x}_{t-1} + \lambda x_{t-1}$ can be chosen by using a grid search in Ref. [0,1] to minimize the RMSE in the training data. It is not unusual for EWMA to provide a competitive RMSE compared to other methods. As an example, **Figure 14** is the daily counts and the EWMA fit for Bay county counts.

Figure 15 is the residuals from the EWMA fit in **Figure 14**.

The standard deviation of the residuals in **Figure 15** is 3.43 while the standard deviation of the counts is 4.43. **Figure 16** is the ACF of the residuals in **Figure 15**.

3.4 Monitoring residuals

One effective option to monitor residuals from a fitted model is Page's statistic applied in this case to residuals in a single county, or to sums of residuals from neighborhoods of counties [17–20]. Reference [20] describes a scan statistic that spans temporally and spatially. Reference [21] illustrates that if the temporal and/or spatial spanning window is selected to provide maximum evidence of an anomaly, then the false alarm probability can be undesirably large unless the variable spanning is taken into account.

To monitor for positive mean shifts due to anomalously-large count(s), Page's statistic applied to the residuals $e_t = x_t - \hat{x}_t$ is defined in Eq. (4) ($S_t = \max(0, S_{t-1} + e_t - k)$). The parameter k is chosen to have good DP for a specific mean shift, but monitoring whether $S_t \geq h$ for threshold h can be effective for a range

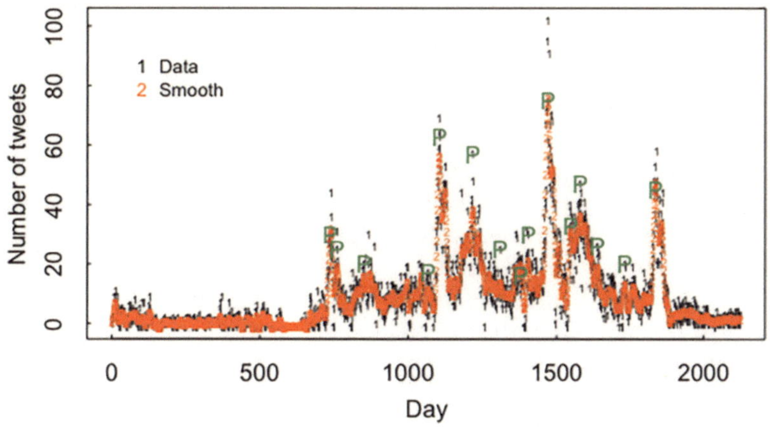

Figure 14.
Daily counts and a smooth fit for bay county. The green "P's" are local maxima in the smooth fit.

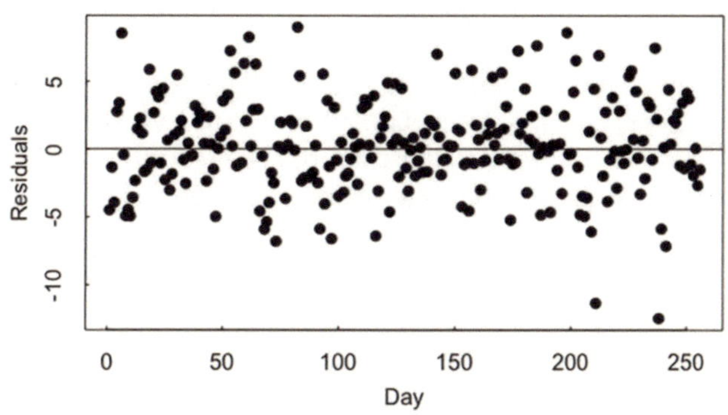

Figure 15.
*Residuals $e = x - \hat{x}$ for the EWMA fit in **Figure 14**.*

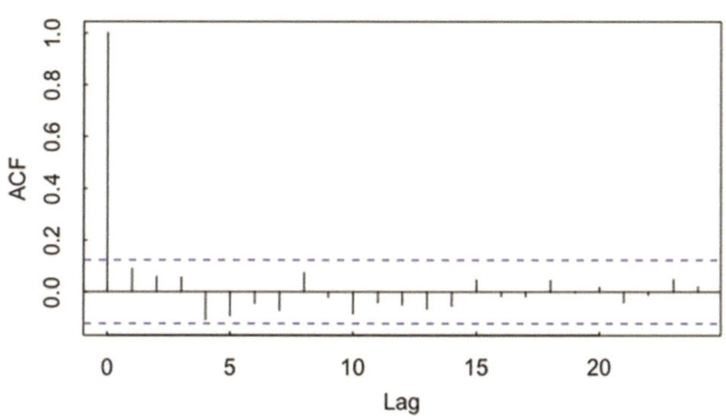

Figure 16.
*ACF of the residuals in **Figure 15**.*

of values of k. For the Poisson distribution, by using the ratio of the likelihood under the background mean μ_B to the shifted mean μ_S, the optimal value of k (leading to the largest DP) is given by $k = \frac{\mu_S - \mu_B}{\ln\left(\frac{\mu_S}{\mu_B}\right)}$. A statistical test based on the maximum of Page's statistic over an analysis window is equivalent to a statistical test based on a scan statistic defined as $\max_{t,j} \sum_{i=0}^{j} \{e_{t+i} - k\}$ over the analysis window.

Because the mean clearly changes over time in the background/training counts, the approach taken here is to fit a model and monitor the residuals e_t that are scaled to have variance 1, so $k = 0.5$ is chosen on the assumption that the residuals are symmetric around 0, and approximately normally distributed, so $k = 0.5$ is optimal (leads to the largest DP) for a mean shift of one standard deviation. However, $k = 0.5$ can lead to large DPs for other mean shift magnitudes. **Figure 17** plots Page's statistic applied to residuals from an EWMA fit for days 1–256 in 2012 in the first four counties (alphabetically: Alachua, Baker, Bay, and Bradford). **Figure 18** is similar to **Figure 17**, but Page's statistic is applied to the net residual in each county plus the residuals in all the nearest-neighbor of each respective county.

To select a threshold (see **Figure 19**) for monitoring all 65 residuals (one from each county) and all 65 residuals (one from each county including the neighboring counties), the 0.95 or 0.99 quantiles of the distribution of the maximum of the 65 or 130 Page's statistic values can be estimated (corresponding to a FAP or either 0.05 or 0.01 per analysis period (256 days in this example).

Figure 20 plots the DP for one-county-at-a-time monitoring and one-county-plus-nearest-neighbors monitoring. The 65 Florida counties are mapped in Appendix 1, defining the neighborhood scheme. For example, the five neighbors of Brevard county are Indian River, Orange, Osceola, Seminole, and Volusia. The nearest-neighbor county DPs are slightly smaller than the one-county-at-a-time DPs because the injected anomalies only impacted one county, and the alarm thresholds are slightly

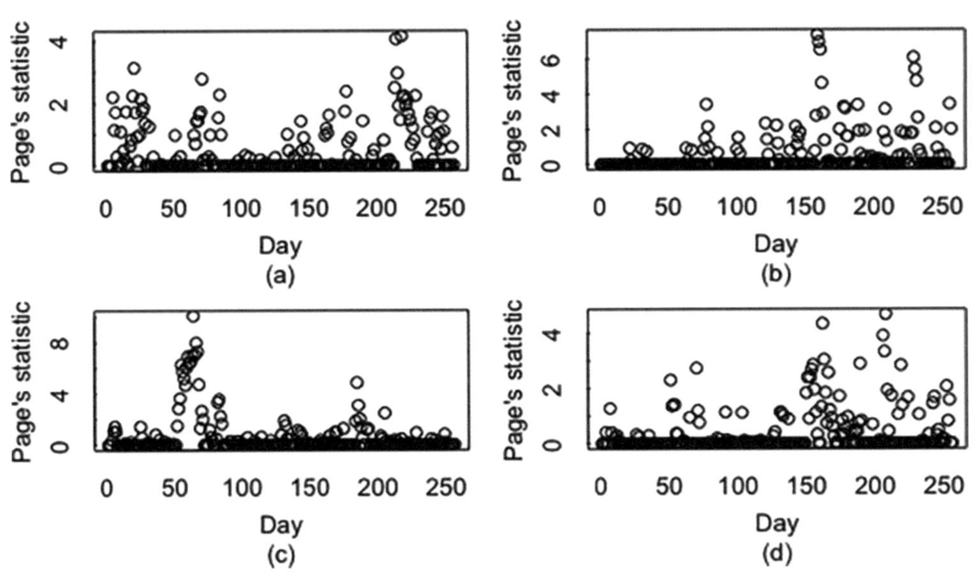

Figure 17.
Page's statistic versus day for Alachua, baker, bay, and Bradford.

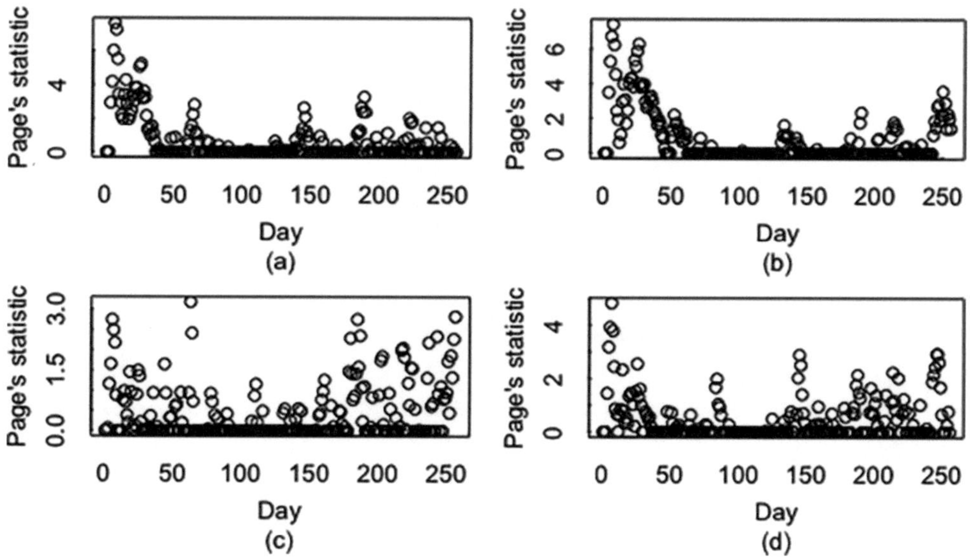

Figure 18.
Page's statistic versus day for Alachua, baker, bay, and Bradford and their respective nearest neighbors.

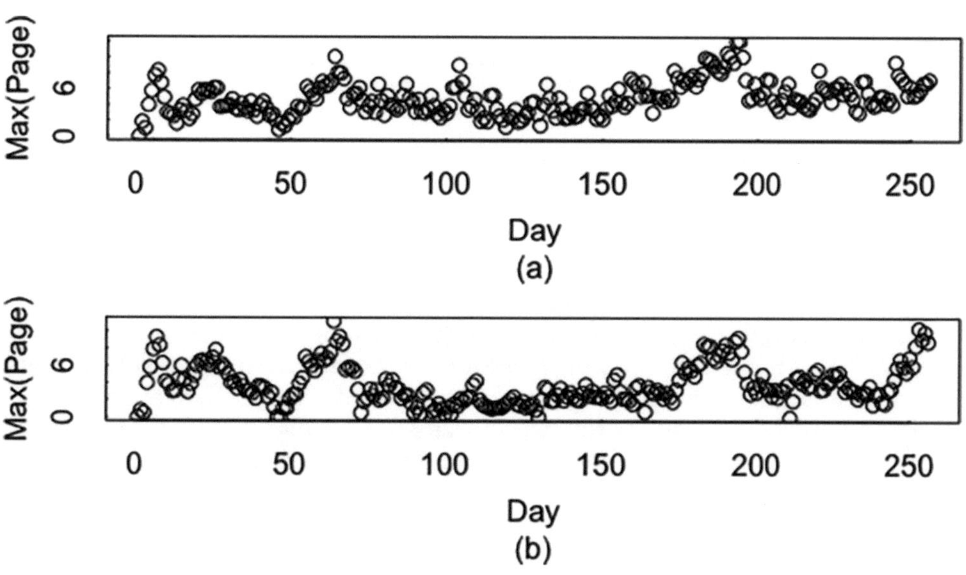

Figure 19.
The maximum value of Page's statistic overall 65 counties (a) and overall 65 counties with residuals from the respective county's neighbors (b).

larger when monitoring both individual counties and individual plus nearest-neighbor counties.

If the injected anomaly impacts all five neighbors of Brevard county then the DPs are much larger than those in **Figure 19**. For example, the DPS are 0.36, 0.78, and 0.99 for a mean shift of 1, 2, or 3 standard deviations in each of the five neighbors.

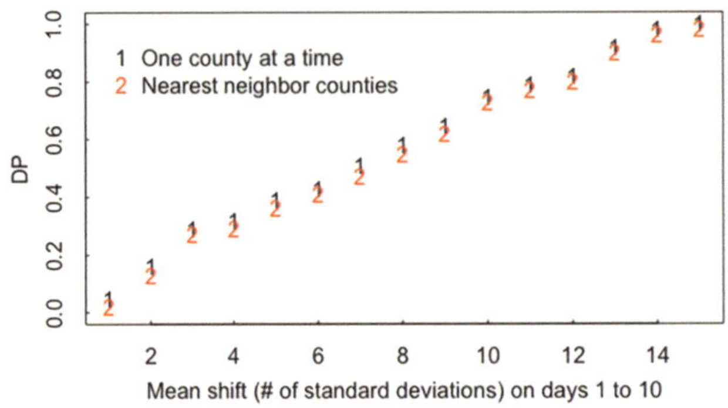

Figure 20.
The DP versus the mean shift occurring on days 1–10 in Brevard county. The nearest-neighbor monitoring includes both one-county-at-a-time and nearest neighbor monitoring.

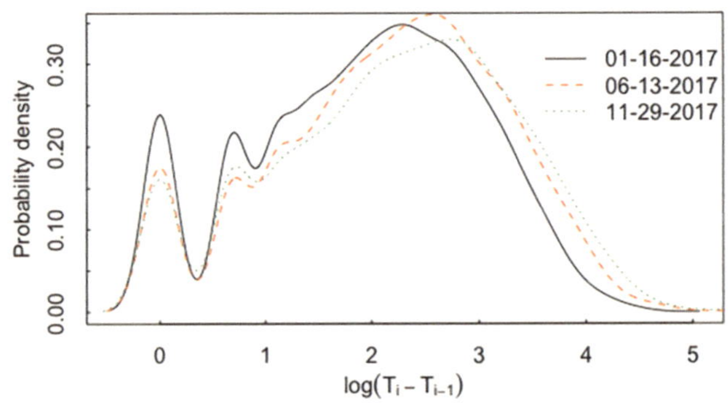

Figure 21.
The probability density of log($T_i - T_{i-1}$*) for each of the 3 days.*

Another Florida data set records each tweet to the nearest second by geo-location in latitude and longitude for each of 3 days during 2017 (January 16, June 13, and November 29). **Figure 21** plots the log of the lag-one time differences, $\log(T_i - T_{i-1})$ for each of the 3 days (with the T values measured in seconds, but the time unit is ignored in applying the log function). If the inter-arrival times followed an exponential distribution with mean $1/\gamma$, then **Figure 20** would exhibit a single peak, and the number of counts in any given time interval of duration t would be distributed as Poisson($\mu = \gamma t$). Clearly, the counts do not exhibit exponential inter-arrival times, and the lag-one time differences between the tweets show considerable similarity across the 3 days. The 3 days have a total of 8102, 6376, and 5839 counts, respectively.

The custom R function generate.data() can be extended to include seasonality such as a single dominant peak per year if appropriate; however, typically more than one local peak is present per year. For example, find peaks applied to the IBR-based smooth finds an average of 4.9, 5.0,3.9, 4.8, 4.1, and 4.7 peaks (across all 65 counties) in years 2010, 2011, ..., 2015, respectively. Simulated data from generate.data() can be used to assess candidate fitting options and to estimate DPs when synthetic anomaly effects are added. However, DP estimates tend to be too optimistic if a data generator such as generate.data() does not include enough realistic effects [18, 19].

Model fitting includes both model selection and estimating parameters in the selected models [13, 18, 19]. Recall that ARMA models are linear models but more generally, for example, AR models such as $x_t = f(x_{t-1}, x_{t-2}, \ldots, x_{t-p}) + e_t$ can be linear in the previous x values or not. Multivariate adaptive regression with splines (MARS) is a flexible nonlinear fitting option that was evaluated using the first 300 days to train and the next 300 days to gest for Brevard county counts. Although the RMSE from MARS fits was 2.48 in training (EWMA has an RMSE of 2.78 in training), the RMSE increased to 3.82 in testing (EWMA has an RMSE of 3.67 in testing), so EWMA remains competitive. Bayesian additive regression trees (BART) are another flexible option to fit AR models. The RMSE for BART (using gbart in R) was 5.07, which is larger than the standard deviation of the daily counts in the testing data (4.04).

Together with the time record of each tweet (to the nearest second), the latitude and longitude provide an option to monitor for spatial and/or temporal clustering. If there is no space–time clustering, then the event of being close in time is independent of the event of being close in time. It is, therefore, possible to check for independence by comparing the number of tweets that are close in space and time to the expected number assuming independence. Arbitrarily defining close in time to be the 0.1 quantiles of all the pairs of time gaps between tweets (and defining close in space to be the 0.1 quantiles of all the spatial distances between pairs of tweets), the 2-by-2 tallies for January 16 is in **Table 1**.

The χ^2test for independence strongly rejects the independence of space and time. Alternatively, the latitude and longitude values can be randomly reordered, breaking any true possible connection between space and time. The resulting test for independence is then not expected to be rejected; however, as an aside, while the random resorting of latitude and longitude reduced the 0.026–0.014, it turns out that 0.014 is large enough to 0.01 in this example, that some type of discretization phenomenon leads to this unexplained behavior. This same discretization phenomenon occurs for other arbitrary definitions of close, such as the 0.05 or 0.2 quantiles instead of the 0.1 quantiles.

Time-tagged Twitter counts with geo-locations are also available for Minnesota, Ohio, and Texas from January 01, 2016 to December 31, 2018 (1093 days with 3 missing days). Reference [19] provides plots of daily Twitter counts for Minnesota, Ohio, and Texas. Decreasing trends are obvious in all three states, and a t-test comparing the first 500 counts to the last 500 counts strongly rejects stationarity.

Reference [19] provides historgrams of the daily counts for Minnesota, Ohio, and Texas, respectively, along with simulated daily counts from a Poisson distribution having that respective state's mean count rate. The real data are much more dispersed than a corresponding Poisson distribution.

	Close in time (0.1 quantile)	Not close in time
Close in space (0.1 quantile)	856,752	23,761,044
Not close in space	2,893,098	5,306,257

Table 1.
*For the 8102 geo-located tweets on January 16, 2017, there are 8102*8101/2 = 32,817,151 comparisons (the sum of the four entries in Table) of time and space. The expected number in the "close in time and space" cell is 0.1 × 0.1 × 32,817,151 = 328171.51, while the observed counts are 856,752 and 856,752/32817151 = 0.026, which is statistically significantly larger than 0.01.*

3.5 Summary

Example 2 and Ref. [19] focused on monitoring residuals from simple EWMA fits. Other possible fitting options, such as MARS or BART for autoregressive modeling are described in Refs. [13, 18, 19]. If there were a more consistent seasonal peak, then model fitting could appropriately include seasonal peak fitting as in Ref. [22] for influenza forecasting. Another data source is Google's flu query data (counts of google searches that seek information about influenza symptoms) as a real-time option to monitor for flu outbreaks [23]. It would be valuable to investigate why the google flu data monitoring option has not been effective. Also, in monitoring for spatial–temporal clustering, it was assumed that "close in space" and "close in time" were defined arbitrarily at the 0.1 quantiles of their respective distributions. If instead several possible quantile values were examined for statistical significance, and the quantile choice leading to the highest evidence of clustering is used, then a simulation-based method [21] could adjust for such maximal selection of statistical evidence of clustering.

4. Example 3: monitoring for nuclear material loss

In nuclear material accounting (NMA), the material balance (MB) is defined as $MB = I_{begin} + T_{in} - T_{out} - I_{end}$, where T_{in} is transferred in; T_{out} is transferred out; I_{begin} is beginning inventory; and I_{end} is ending inventory [24–34]. All terms involve measured material, so the MB values should vary around 0 if there is no NM loss. The measurement error standard deviation of the MB is denoted σ_{MB}. Typically, many measurements are combined to estimate the terms T_{in}, I_{begin}, T_{out}, and I_{end} in the MB; therefore, the central limit effect and years of experience suggests that MBs in most facilities will be approximately normally distributed with a mean equal to the true NM loss μ and standard deviation σ_{MB}, which is expressed as $X \sim N(\mu,\sigma_{MB})$, where X denotes the MB. If the MB at a given time ("balance period") exceeds $k\ \sigma_{MB}$ with k in the 2–3 range, then the NMA system "alarms."

A sequence of n MBs is often assumed to have approximately a multivariate normal distribution $X = X_1, \ldots, X_n \sim N(\mu, \Sigma)$, where the n-by-n covariance matrix

$$\Sigma = \begin{pmatrix} \sigma_1^2 \sigma_{12}^2 \ldots \sigma_{1n}^2 \\ \sigma_{21}^2 \sigma_2^2 \ldots \sigma_{2n}^2 \\ \ldots \\ \sigma_{n1}^2 \sigma_{n2}^2 \ldots \sigma_n^2 \end{pmatrix}$$

. Estimating Σ is often one of the most tedious steps in frequent NMA (near real-time accounting, NRTA). A simplified example of estimating a component of Σ using a model of a generic electrochemical facility with one input stream, one output stream, and two inventory items is as follows. First, each individual measurement method is modeled with a measurement error model. A typical model for multiplicative errors is $M_{ij} = T_i(1 + S_i + R_{ij})$ with $S_i \sim N(0, \delta_S^2)$ and $R_{ij} \sim N(0, \delta_R^2)$, where the jth measurement results (often $j = 1$) M_{ij} of item i, T_i is the true value of item i, R_{ij} is a random error of item i, S_i is a short-term systematic error for item i. Then, the error variance for the two inventory items is $\sigma_I^2 = (T_1 + T_2)^2\delta_S^2 + T_1^2\delta_R^2 + T_2^2\delta_R^2$, just as one example [26–29, 34] of error modeling and variance propagation used to estimate a component of Σ.

In the early 1980s, some believed that a plant reporting an MB every 30 days would have a larger detection probability (DP) than that same plant reporting an MB every year (typically a facility is inventoried and cleanout out approximately once per year). However, Ref. [27] then showed that for optimal (from the diverter's viewpoint, meaning that the DP is minimized) protracted diversion with the per-period loss being proportional to the row sums of the covariance matrix, Σ of the MB series, annual MB testing has larger DP than monthly mass balance testing. Reference [27] dampened hopes that frequent NMA, referred to as NRTA, would allow challenging diversion DP goals to be met. However, Ref. [27] conceded that NRTA has shorter detection times and higher DPs against abrupt diversion. Reference [27] showed that the best statistical test, the Neyman-Pearson (NP)-based matched filter for the worse case loss with $\mu_i^* = \frac{K\sum_{j=1}^{n}\Sigma_{i,j}}{\sum_{i=1}^{n}\sum_{j=1}^{b}\Sigma_{i,j}}$, is based on the cumsum, $C_t = \sum_{i=1}^{t} x_i$.

There are several reasons to apply statistical tests to a transformed sequence defined as $Y_i = \{X_i - E(X_i|X_{i-1}, X_{i-2}, \dots, X_1)\}/\tilde{\sigma}_i$ where E denotes the expectation and the standard deviation $\tilde{\sigma}_i$ of $\{X_i - E(X_i|X_{i-1}, X_{i-2}, \dots, X_1$ is $\tilde{\sigma}_i = \sqrt{\sigma_{ii}^2 - f\Sigma^{-1}f^T}$ where $f = \Sigma_{i,1(i-1)}$, the 1 to (i-1) entries in the ith row of Σ [2, 7]. From properties of the MVN, each component Y_i vector can be computed by calculating the conditional mean $E(X_i|X_{i-1}, X_{i-2}, \dots, X_1) = f\Sigma_{i-1}^{-1}X_{i-1}$ where Σ_{i-1}^{-1} is the inverse of the (i-1)-by-(i-1) matrix Σ_{i-1} that corresponds to balance period 1 through period i-1. The Cholesky factorization $\Sigma = LU$ leads to a more computationally efficient recursive approach that avoids matrix inversion [28]. The transformed sequence $Y = L^{-1}$ X has $\Sigma_Y = I$, so Y is a residual time series.

This is a logistic advantage and a DP advantage to transform the $X_1, X_2, \dots, X_n$ time series to the series of residuals $Y_1, Y_2, \dots, Y_n$ known in NMA as the SITMUF (standardized, independently transformed material unaccounted for sequence, here MUF is another term for the MB). The logistic advantage is that the SITMUF time series is iid N(0,1) if the loss $\mu = 0$, so alarm thresholds depend only on the sequence length n and the desired FAP. The DP advantage is a DP increase for many loss vectors arises because the variance of the SITMUF sequence decreases over time, so particularly if a diversion occurs late in the analysis period, the DP is larger for the Y sequence than for the X sequence. Note that one cannot claim higher DP for the Y sequence than for the X sequence in general, because the true loss scenario is never known, and the DP can be larger for X than for Y for some loss scenarios. Modern NRTA systems use a suite of several statistical tests, usually applied to the Y series.

Example DPs are plotted in **Figures 22–24** (three different loss vectors for n = 16) for these five tests: (1) SITMUF test, (2) Page's test applied to the SITMUF series, (3) CUMSUM test, (4) a combination of (1–3), and (5) the NP-based matched filter is also useful to compute the largest possible DP for a specified loss. The alarm threshold h is chosen so that the FAP per analysis period (usually one year) is 0.05 or whatever FAP is specified. In **Figures 22–24**, the covariance matrix Σ from Ref. [34] has 1 on the diagonal, −0.48 on the lag-one off diagonals, and 0.01 on all higher-lag off diagonals. The loss 1 vector (**Figure 21**) is 0 on periods 1 to 5, constant on periods 6–10, then 0 on periods 11–16 (the nonzero entries summing to the quantity plotted on the horizontal axis). The loss 2 vector (**Figure 22**) is all 0 except for an abrupt loss on period 6. The loss 3 vector (**Figure 23**) is constant for all 16 periods.

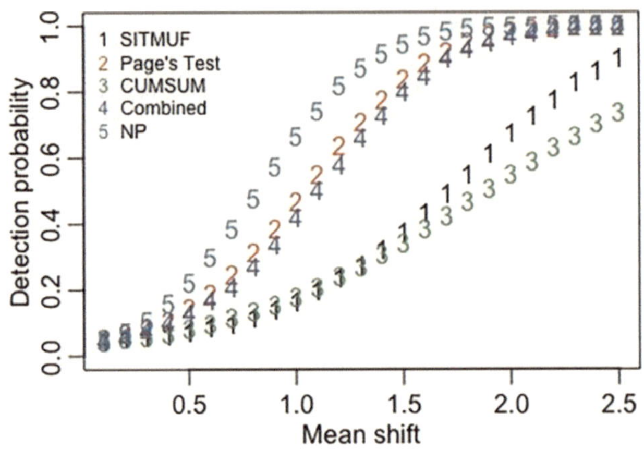

Figure 22.
DP versus the total mean shift (loss) for n = 16, *for loss vector 1 (constant loss on periods 6 to 10).*

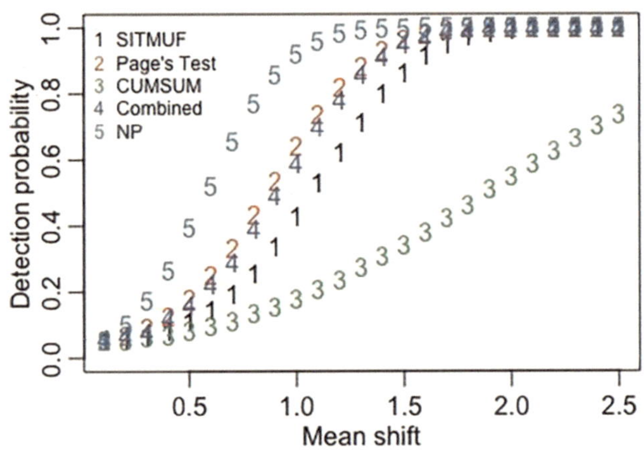

Figure 23.
DP versus the total mean shift (loss) for n = 16, *for loss vector 2 (abrupt loss on period 6).*

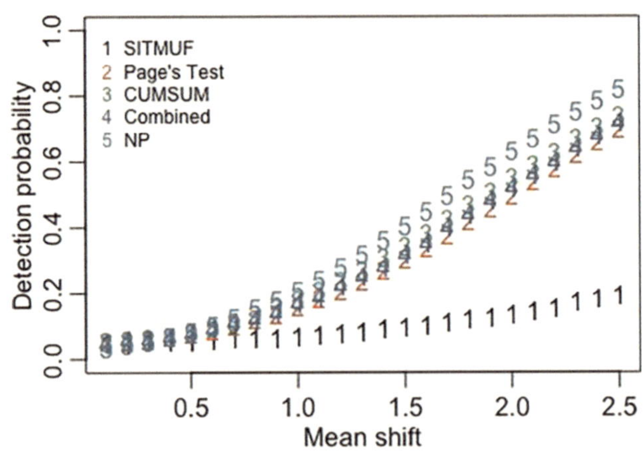

Figure 24.
DP versus the total mean shift (loss) for n = 16, *for loss vector 3 (constant loss).*

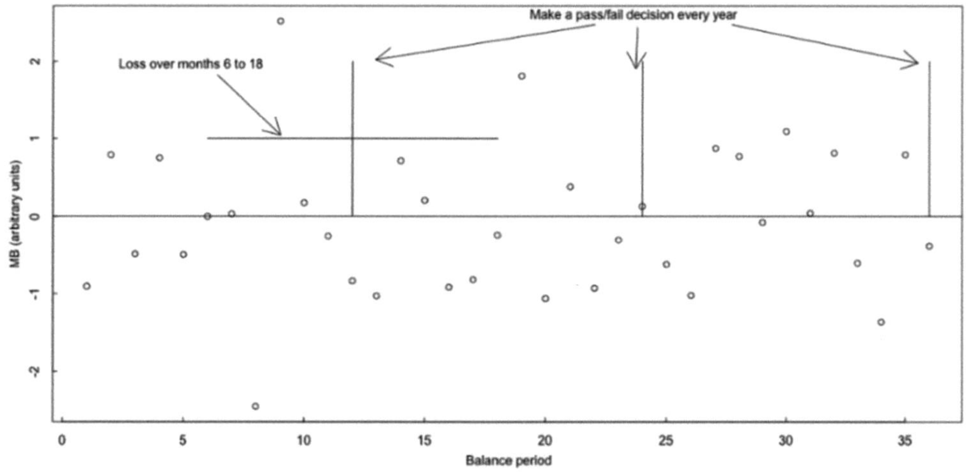

Figure 25.
MB sequences over 36 months using fixed-period (annual) decision periods.

This example is concluded with two remarks.

Remark 1. Reference [24] showed that assuming the model $X_1, \ldots, X_n \sim N(\mu, \Sigma)$ leads to larger DPs than fitting an ARMA model on training data for which it would have to be assumed that the NM loss $\mu = 0$ [31], provided Σ is well estimated. If Σ is not well estimated, a Bayesian updating scheme to improve the estimate of Σ could be used on training data for which the NM loss $\mu = 0$ [35].

Remark 2. Figure 25 illustrates fixed-period testing and data-driven testing [29]. Some versions of NRTA use the most recent 1-year length sequences, so for $n = 12$ balance periods per year, the first evaluation period is months 1–12, the second evaluation period is months 2–13, etc. This scheme allows for a statistical decision to be made at every annual physical inventory, such as at months 12, 24, and 36. An alternate scheme consisting of a hybrid of period-driven and data-driven testing is described in Ref. [29], where it is pointed out that one should not simply truncate sequential statistical tests at the time of the annual physical inventory because the adversary could remove a portion of an SQ during year 1 and the remaining portion during year 2. The scan statistic has the highest DP (0.95 in this case, verified by simulation) if one knows that a loss will occur over a 12-month period with an unknown start period (such as month 7). The scan statistic computes a moving sum of months 1–12, 2–13, 3–14, etc.

5. Summary

This chapter has described three change-detection examples. Example 1 monitored for patterns of large residuals, such as a consecutive string of three residuals exceeding a threshold. Example 2 monitored for excessive numbers of tweets in any of the 65 Florida counties. Example 3 monitored for nuclear material loss. Portions of examples 1 and 3 have been published as cited in Refs. [5, 34]. Example 2 is entirely new. Page's statistic is generally recommended because of its reasonably large DP for a range of change patterns, such as any of those in **Figure 1**.

Appendix 1: The Florida counties in example 2

> ctylist
[1] "alachua" "baker" "bay" "bradford"
[5] "brevard" "broward" "calhoun" "charlotte"
[9] "citrus" "clay" "collier" "columbia"
[13] "desoto" "dixie" "duval" "escambia"
[17] "flagler" "franklin" "gadsden" "gilchrist"
[21] "glades" "gulf" "hamilton" "hardee"
[25] "hendry" "hernando" "highlands" "hillsborough"
[29] "holmes" "indian river" "jackson" "jefferson"
[33] "lake" "lee" "leon" "levy"
[37] "madison" "manatee" "marion" "martin"
[41] "miami-dade" "modroe" "nassau" "okaloosa"
[45] "okeechobee" "orange" "osceola" "palm beach"
[49] "pasco" "pinellas" "polk" "putnam"
[53] "st. johns" "st. lucie" "santa rosa" "sarasota"
[57] "seminole" "sumter" "sumannee" "taylor"
[61] "union" "volusia" "wakulla" "walton"
[65] "washington".

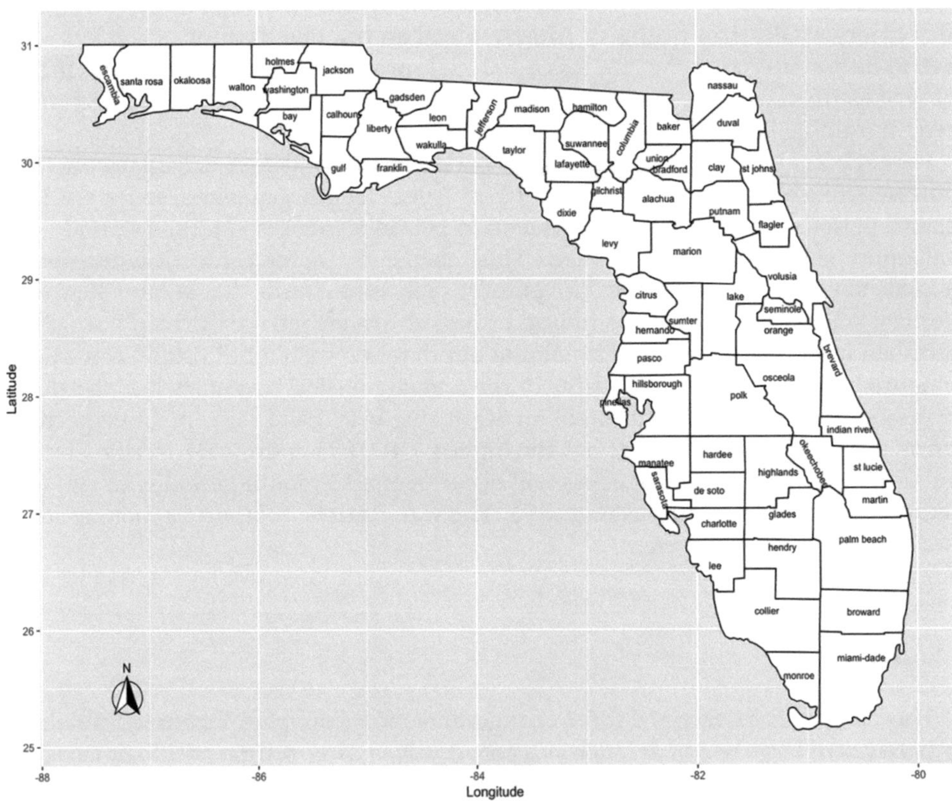

Figure A.1.
The 65 Florida countries. In the available data, liberty county is merged into Gadsden and Lafayatte county is merged into Madison to reduce the 67 Florida counties to 65. Therefore, there are 65 identified regions (counties) for which spatial and/or temporal residuals patterns can be monitored for change.

Figure A.1 is the 67 Florida counties, and this data merged Liberty into Gadsden and merged Lafayatte into Madison to reduce the 67 Florida counties to 65.

```
> ctynbrs[5,] # Brevard county: neighbors of Brevard county
[1] 0 0 0 0 0 0 0 0 0 0 0 0 0 0 0 0 0 0 0 0 0 0 0 0 0 0 0 0 1 0 0 0 0 0 0 0 0
[38] 0 0 0 0 0 0 0 0 1 1 0 0 0 0 0 0 0 0 0 0 0 1 0 0 0 0 1 0 0 0.
ctylist[ctynbrs[5,]
> ctylist[as.logical(ctynbrs[5,])]
[1] "Indian river"  "orange"  "osceola"  "seminole"  "volusia"
```

Author details

Tom Burr* and Kim Kaufeld
Los Alamos National Laboratory, Los Alamos USA

*Address all correspondence to: tburr@lanl.gov

References

[1] Chatfield C. The Analysis of Time Series: An Introduction. 6th ed. London, United Kingdom: Chapman and Hall; 2004

[2] Shumway R, Stoffer D. Time Series Analysis and its Applications with R Examples. 4th ed. Pittsburgh: Springer; 2016

[3] Lucas J. Counted data Cusums. Technometrics. 1985;**27**(2):129-144

[4] R Core Team. R. A Language and Environment for Statistical Computing. R Foundation for Statistical Computing. Vienna, Austria. Available from: https://www.R-project.org/: R Foundation for Statistical Computing; 2017

[5] Burr T, Henderson B. Scanning for clusters of large values in time series: Application of the Stein-Chen method. Applied Mathematics. 2021;**12**:1031-1037

[6] Arratia R, Goldstein L, Gordon L. Poisson approximation and the Chen-Stein methods. Statistical Science. 1990; **5**(4):403-434

[7] Sahatsathatsana C. Applications of the Stein-Chen method for the problem of coincidences. International Journal of Pure and Applied Mathematics. 2017; **116**(1):49-59

[8] Kim S. A use of the Stein-Chen method in time series analysis. Journal of Applied Probability. 2000;**37**(4):1129-1136

[9] Aleksandrov B, Weis C, Jentsch C. Goodness-of-fit tests for Poisson count time series based of the Stein-Chen identity. Statistica Neerlandica. 2021;**76**: 35-64

[10] Weis C, Aleksandrov B. Computing bivariate Poisson moments using stein-Chen identities. The American Statistician. 2022;**76**(1):10-15

[11] Borror C, Champ E, Rigdon S. Poisson EWMA control charts. Journal of Quality Technology. 2018;**30**(4):352-361. DOI: 10.1080/00224065.1998.11979871

[12] Venables W, Ripley B. Modern Applied Statistics with S-Plus. New York: Springer; 1999

[13] Hastie T, Tibshiranii R, Friedman J. Elements of Statistical Learning. New York: Springer; 2001

[14] Burr T, Hengartner N, Matzner-Løber E, Myers S, Rouviere L. Smoothing low resolution gamma spectra. IEEE Transactions on Nuclear Science. 2010;**57**:2831-2840

[15] Cornillon P, Hengartner N, Jegou N, Matzner-Løber E. Iterative bias reduction: A comparative study. Statistics and Computing. 2013;**23**(6):777-791

[16] Hengartner N, Matzner-Lober E, Rouviere L, Burr T. Multiplicative Bias Corrected Nonparametric Smoothers with Application to Nuclear Energy Spectrum Estimation, Nonparametric Statistics. 3rd ISNPS ed. Avignon, France: Springer; 2016 arXiv Preprint arXiv:0908.0128

[17] Mathes R, Lall R, Levin-Rector A, Sell J, Paladini M, Konty K, et al. Evaluating and implementing temporal, spatial, and spatio-temporal methods for outbreak detection in a local syndromic surveillance system. PLoS ONE. 2017;**12** (9):e0184419. DOI: 10.1371/journal.pone.0184419

[18] Burr T, Graves T, Klaman R, Michalek S, Picard R, Hengartner N.

Accounting for seasonal patterns in syndromic surveillance data for outbreak detection, BioMedCentral. Medical Informatics and Decision Making. 2006; **6**:40

[19] Burr T, Kaufeld K. Statistical Evaluation of Daily Tweet Counts from Florida, Minnesota, Ohio, and Texas. New Mexico, United States: Los Alamos National Laboratory Report; 2021

[20] Kulldorff M. Prospective time periodic geographical disease surveillance using a scan statistic. Journal of the Royal Statistical Society. 2001; **A164**:61-72

[21] Burr T. Maximally selected measures of evidence of disease clustering. Statistics in Medicine. 2001;**20**: 1443-1460

[22] Osthus D, Moran K. Multiscale influenza forecasting, nature communications 12. Art. 2021;**2991**

[23] Pervais F, Pervaiz M, Rehman N, Saif U. FluBreaks early epidemic detection for google flu trends. Journal of Medical Internet Research. 2012;**14**(5): e125. DOI: 10/2196jmir.2002

[24] Burr T, Hamada MS. Smoothing and time series modeling of nuclear material accounting data for protracted diversion detection. Nuclear Science and Engineering. 2014;**177**:307-320

[25] Burr T, Hamada MS. Statistical Challenges in Integrated Nuclear Safeguards, Nuclear Science in the Series Energy Science and Technology. Vol. 4 (12). Vienna, Austria: IAEA; 2014

[26] Burr T, Hamada MS. Revisiting Statistical Aspects of Nuclear Material Accounting Science and Technology of Nuclear Installations. London, United Kingdom: Hindawi Publishing Corporation; 2013. pp. 1-15. DOI: 10.1155/2013/961360

[27] Avenhaus R, Jaech J. On subdividing material balances in time and/or space. Journal of Nuclear Materials Management. 1981;**10**:24-34

[28] Picard R. Sequential analysis of material balances. Journal of Nuclear Materials Management. 1987;**15**(2):38-42

[29] Burr T, Hamada MS, Ticknor L, Sprinkle J. Hybrid statistical testing for nuclear material accounting data and/or process monitoring data in nuclear safeguards. Energies. 2015;**8**:501-528

[30] Prasad S, Booth T, Hu M, Deligonul S. The detection of nuclear materials losses. Decision Sciences. 2007; **26**(2):265-281

[31] Speed T, Culpin D. The role of statistics in nuclear materials accounting: Issues and problems. Journal of the Royal Statistical Society A. 1986;**149**(4): 281-313

[32] Downing D, Pike D, Morrison G. Analysis of MUF data using ARlMA models. Journal of Nuclear Material Management. 1978;**7**(4):80-86

[33] Bonner E, Burr T, Krieger T, Martin K, Norman C. Comprehensive Uncertainty Quantification in Nuclear Safeguards, Science and Technology of Nuclear Installations. London, United Kingdom: Hindawi Publishing Corporation; 2017. pp. 1-16. DOI: 10.1155/2017/2679243

[34] Burr T, Hamada MS. Bayesian updating of material balances covariance matrices using training data. International Journal of Prognostics and Health Monitoring. 2014;**5**(1):006-013

Section 7

Forecasting and Prediction

Chapter 7

Comparison of the Out-of-Sample Forecast for Inflation Rates in Nigeria Using ARIMA and ARIMAX Models

Monday Osagie Adenomon and Felicia Oshuwalle Madu

Abstract

This book chapter compares the out-of-sample forecast for inflation rates in Nigeria using ARIMAX and ARIMA models. To achieve this, Annual Data on Exchange Rate, Inflation Rate, Interest Rate and Unemployment Rate from 1981 to 2017 was sourced from Central Bank of Nigeria (CBN). The analysis used data from 1981 to 2010 while 2011 to 2017 was used to valid the forecast from the ARIMA and ARIMAX models. The preliminary analysis revealed that natural log transform of inflation rate is normally distributed and stationary at first difference while Exchange Rate, Inflation Rate, Interest Rate and Unemployment Rate were used as exogenous variables in the ARIMAX models. The following models ARIMA(1,1,0), ARIMA(1,1,1), ARIMA (0,1,1), ARIMAX(1,1,0), ARIMAX(1,1,1) and ARIMAX(0,1,1) were compared for both in-sample and out-of-sample forecasts. Using the Root Mean Square Error (RMSE) as selection criteria, ARIMAX(0,1,1) with RMSE of 0.6810 emerged as superior model for the in-sample forecast for forecasting inflation rate in Nigeria while ARIMA(1,1,1) emerged as a superior model for the out-of-sample forecast for inflation rate in Nigeria and its forecast for inflation revealed a negative growth in inflation in Nigeria. This study therefore recommended ARIMA(1,1,1) model be used for out-of-sample forecast for inflation rate in Nigeria.

Keywords: forecasting, inflation, ARIMA, ARIMAX, RMSE

1. Introduction

A time series can be considered as an ordered sequence of observations, of which the ordering is through time [1]. The ordering could be equally spaced time interval or may take other dimensions, such as space [2]. The applications of time series can be found in engineering, geophysics, business, economics, medical studies, meteorology, quality control, social sciences and agriculture. The list of the areas cannot be exhausted.

There are various objectives for studying time series. These include the understanding and description of the generated mechanism, the forecasting of future values and optimum control of a system. The uses of time series analysis are (i). It helps in the analysis of past behavior of a variable, (ii) it helps in forecasting (iii). It helps in evaluation of current achievement (iv). It helps in making comparative studied. Therefore, the body of statistical methodology available for analyzing time series is referred to as time series analysis [3].

Univariate time series modeling is very useful in forecasting such series. In the class of univariate time series models, the model proposed by Box and Jenkins [4] as Autoregressive Moving Average (ARMA) and Autoregressive Integrated Moving Average (ARIMA) models are most popular and excellent while Autoregressive Integrated Moving Average with Explanatory Variable (ARIMAX) is becoming also popular because researchers have found that ARIMAX model can outperformed the ARMA or ARIMA models [5]. These models are applied in almost all fields of endeavors such as engineering, geophysics, business, economics, finance, agriculture, medical sciences, social sciences, meteorology, quality control etc. [1]. This chapter considered forecasting inflation using Exchange, Interest and unemployment rates as Exogenous variable with the of ARIMAX model.

Inflation, exchange, interest, unemployment, and growth rates are the big macroeconomic issues of our time [6]. Inflation is bad, especially when unexpected, because it distorts the working of the price system, creates arbitrary redistribution from debtors to creditors, creates incentives for speculative as opposed to productive investment activity, and is usually costly to eliminate. Inflation can be defined as a positive rate of growth of the general price level. Eitrheim et al. [7] noted that Inflation, exchange, interest, unemployment, and growth rates can affect any economy (either positive or negative) that is why these macroeconomic variables are of great interest to Central Banks of many countries of the world.

Therefore the chapter considered forecasting inflation in Nigeria using ARIMA and ARIMAX Models.

2. Empirical literature reviews of previous studies

Inflation may be defined as a positive rate of growth of the general price level. Eitrheim et al. [7] noted that Inflation, exchange, interest, unemployment, and growth rates can affect any economy (either positive or negative) that is why these macroeconomic variables are of great interest to Central Banks of many countries of the world.

Several authors have studied the influence of inflation rates on other macroeconomic variable and how other macroeconomic variable affect inflation. Some of the authors include: Omotor [8] who studied the relationship between inflation and stock market returns in Nigeria; Shittu and Yaya, [9] studied the inflation rates in Nigeria, United States and United Kingdom using fractionally integrated logistic smooth transitions in time series; Abraham [10] studied the short and long runs effect of inflation rates on All Share Index (ASI) in Nigeria; Musa and Gulumbe, [11] studied the interrelationship between inflation rate and government revenues in Nigeria using Autoregressive Distributed Lag (ARDL) model. Other economy have been also studied by several authors. Such authors include: Furuoka [12] who studied the interrelation between unemployment and inflation in the Philipines using Vector Error Correction Model (VECM). Omar and Sarkar [13] studied the relationship between commodity prices and exchange rate in the light of global financial crisis in Australia using Vector Error Correction Model (VECM). Mohaddes and Raissi [14] examined the long-run relationship between consumer price index industrial workers (CPI-IW) inflation and GDP growth in India using cross-sectional Augmented distributed lag

(CS-DL) as well as standard panel ARDL method. The findings of Mohaddes and Raissi suggested that, on the average, there is a negative long-run relationship between inflation and economic growth in India.

Mida [15] revealed that changes in inflation rate have the opposite effects on the exchange rate that is a rising inflation rate can depreciate the exchange rate.

Nastansky and Strohe [16] analyzed the interaction between inflation rate and public debt in Germany using quarterly data from 1991 to 2014 using Vector Error Correction Model (VECM). Their result revealed a strong positive relationship between inflation rate and public debt.

Gillitzer [17] empirically assessed the performance of the Sticky Information Phillips Curve (SIPC) for Australia. The study revealed that the estimates were sensitive to inflation measures and sample period. Also poor performance of the SIPC revealed the fact that inflation can be deficit to the model.

The following are empirical literature of the application of ARIMA, ARIMAX and Other Time Series models:

Kongcharoen and Kruangpradit [5] examined and forecast Thailand exports to major trade partners using ARIMA and ARIMAX models. They found that ARIMAX outperforms the ARIMA Model.

Stock and Watson [18] empirically found out that time series regression model that includes leading indicators into the model improves forecast performance.

Bougas [19] examined the Canadian Air transport sectors divided into domestic, transboarder (US) and International flights using various time series forecasting models namely: Harmonic regression, Holt-Winters Exponential smoothing, ARIMA and SARIMA Regressions. The result indicated that all models provide accurate forecast with MAPE and RMSE scores below 10% on the average.

Adenomon and Tela [20] fitted and forecasted inflation rates in Nigeria for annual data covering 1970 to 2014. Among the ARIMA competing models, ARIMA (1,1,2) was superior. While forecast for inflation rates revealed a negative trend.

Styrvold and Nereng [21] compared ARIMA model with classical regression model and VAR to model real rental rates as a function of previous periods' rate, employment rates, real interest rates and vacancy rates. The studied concluded that classical linear regression model is able to produce the most precise forecasts, although the precision is not satisfactory.

Amadeh et al. [22] modeled and predicted the Persian Gulf Gas-Oil F. O. B using ARIMA and ARFIMA models on weekly data of gas-oil prices. Their results revealed that ARFIMA model performed better than ARIMA.

Avuglar et al. [23] applied ARIMA time series model to accident data from 1991 to 2011 in Ghana. They recommended ARIMA (0,2,1) as the best model.

Moshiri and Foroutan [24] modeled and forecast daily crude oil future prices from 1983 to 2003 listed in NYMEX using ARIMA and GARCH models. They further improved forecast with the use of Neural network models.

Adenomon [2] modeled and forecasted the evolution of unemployment rates in Nigeria using ARIMA model on annual data for the period of 1972 to 2014. The study revealed ARIMA (2,1,2) as superior model for unemployment rates in Nigeria.

3. Model specification

This section considered the models used in this chapter.

3.1 Unit root test

Engle and Granger [25] considered seven test statistics in a simulation study to test cointegration. They concluded that the Augmented Dickey Fuller test was recommended and can be used as a rough guide in applied work. The essence of the unit root test is to avoid spurious regression.

To identify a unit root, we can run the regression

$$\Delta Y_t = b_o + \sum_{j=1}^{k} b_j \Delta Y_{t-j} + \beta t + \gamma Y_{t-1} + u_t \tag{1}$$

The model above can be run without t if a time trend is not necessary [26]. If unit root exist, differencing of Y will result in a white-noise series (that is no correlation with Y_{t-1}).

The null hypothesis of no unit root test in the Augmented Dickey-Fuller (ADF) test is given as H_o: $\beta = \gamma = 0$ (if trend is consider, we use F-test) and H_o: $\gamma = 0$ (if there is no trend is consider, we use t-test). If the null hypothesis is not rejected, this suggest that unit root exist and the differencing of the data is required before running a regression. When the null hypothesis is rejected, the data are refer to as stationary and it can be analyzed without any form of differencing [27].

3.2 ARIMA model and estimation

ARIMA model can be viewed as an approach that combines the moving average and the autoregressive models [28]. Box and Jenkins are the pioneers of the ARIMA model that is why it is refer to as the Box-Jenkins (BJ) methodology, but in time series literature is known as the ARIMA methodology [29]. The ARIMA models allow Y_t to be explained by the past, or lagged, values of Y_t and stochastic error terms.

The ARMA (p, q) model is a combination of the AR and MA model which is given as

$$y_t = a_o + a_1 y_{t-1} a_2 y_{t-2} + \ldots + a_p y_{t-p} - b_1 u_{t-1} - b_2 u_{t-2} - \ldots - b_q u_{t-q} + u_t \tag{2}$$

Box and Jenkins recommend difference non-stationary series one or more times to achieve stationarity. Doing so produces an ARIMA model, with the 'I' standing for 'Integrated'. But its first difference $\Delta y_t = y_t - y_{t-1} = u_t$ is stationary, so y is 'Integrated of order 1' or $y \sim I(1)$.

The primary stages in building a Box-Jenkins time series model are model identification; model estimation and model validation. The Theoretical features of autocorrelation function (ACF) and partial autocorrelation function (PACF) (**Table 1**).

Type of model	Typical feature of ACF	Typical feature of PACF
AR(p)	It decays exponentially or with damped sine wave pattern or both	Significant spikes are seen through lags p
MA(q)	Significant spikes are seen through lags p	It declines exponentially
ARMA(p,q)	It exponentially decay	It exponentially decay

Table 1.
Theoretical features of autocorrelation function (ACF) and partial autocorrelation function (PACF).

After fitting ARIMA Model, test for adequacy of the fitted model (the chi-squared test for goodness of fit) called Ljung-Box test [30] is required. The Ljung-Box test is based on all the residual ACF as a set. The test statistic is as follows $Q = n(n+2)\sum_{i=1}^{k}(n-i)^{-1}\gamma_i^2(\hat{a})$ where $\gamma_i^2(\hat{a})$ is the estimate for $\rho_j(\hat{a})$ and n is the number of observations used to estimate the model. The statistic Q follows approximately the chi-squared distribution with k-v degrees of freedom, where v is the number of parameters estimated in the model. If we do not reject the null hypothesis, then it implies that the fitted model is adequate.

3.3 ARIMAX model

The ARIMA model will be extended into ARIMA model with explanatory variable (X_t), called ARIMAX(p,d,q). Specifically, ARIMAX(p,d,q) can be represented by

$$\varphi(L)(1-L)^d Y_t = \Theta(L)X_t + \theta(L)\varepsilon_t \tag{3}$$

Where L is the lag operator, d = difference order, p is the AR order, q is the MA order, explanatory variables (X_t) and ε_t is the error term while φ,Θ,θ are the coefficients of the AR, MA and the exogenous variables [5].

3.4 Forecast assessment criteria

We considered the following forecast assessment criteria in this book chapter:

1. Mean Absolute Error (MAE) is given as $MAE_j = \frac{\sum_{i=1}^{n}|e_i|}{n}$. This statistic measures the deviation from the series in its absolute terms, and measures the forecast bias. The MAE is one of the most common ones used for analyzing the quality of different forecasts.

2. Root Mean Square Error (RMSE) is given as $RMSE_j = \sqrt{\frac{\sum_{i}^{n}(y_i - y^f)^2}{n}}$ where y_i is the time series data and y^f is the forecast value of y [31].

For the two measures above, the smaller the value, the better the fit of the model [3].

4. Method of data collection

The data used in this book chapter was collected from a secondary source. Annual Data on Exchange Rate, Inflation Rate, Interest Rate, Unemployment Rate from 1981 to 2017 was sourced from Central Bank of Nigeria (CBN) Statistical Bulletin [32]. Inflation rate is the variable of interest (response variable) while the exogenous variables are Exchange Rate, Interest Rate, Unemployment Rate. The variables are transformed using the natural logarithm to ensures stability and normality, and to reduce skewness and variability. Also the analysis was used data from 1981 to 2010 while 2011 to 2017 was used to valid the forecast from the ARIMA and ARIMAX models.

5. Results and discussion

The section presented the results emanating from the analysis and discussions of results. The data analysis of this book chapter was carried out in R software environment using tseries and TSA packages.

Figure 1 below shows the inflation rate in Nigeria from 1981 to 2010. It is observed that Nigeria experienced inflation from 1993 to 1996. While the inflation rates were low from 2000 to 2010.

Figure 2 below shows the natural log transform of inflation rate in Nigeria from 1981 to 2010. It is observed that Nigeria experienced inflation from 1993 to 1996. While the inflation rates were low from 2000 to 2010. In addition there is reduction in the trend of inflation rates after transformation.

Figure 3 below presented the plots of Interest Rate (INT), Unemployment Rate (UNE) and Exchange Rate (EX) from 1981 to 2010 in Nigeria. The interest rate shows some decrease 2002 to 2010 but for unemployment rates and exchange rates shows an increase from 2002 to 2009. This situation about unemployment and exchange rates will definitely affect the standard of living in Nigeria if not properly control.

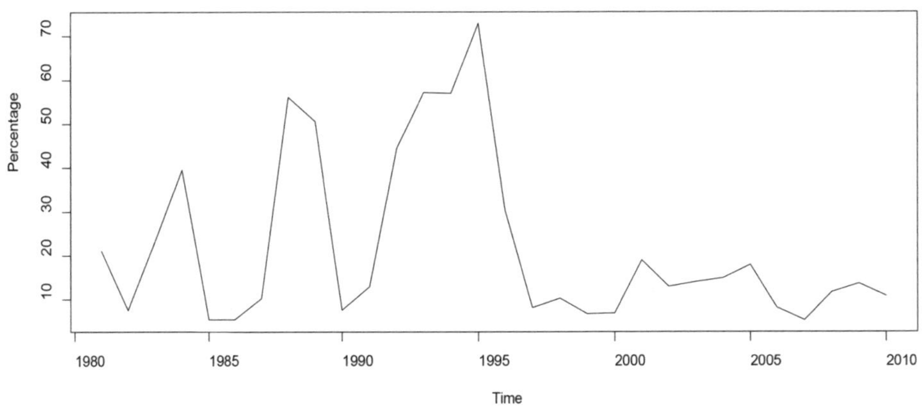

Figure 1.
Plot of Inflation Rate in Nigeria from 1981 to 2010.

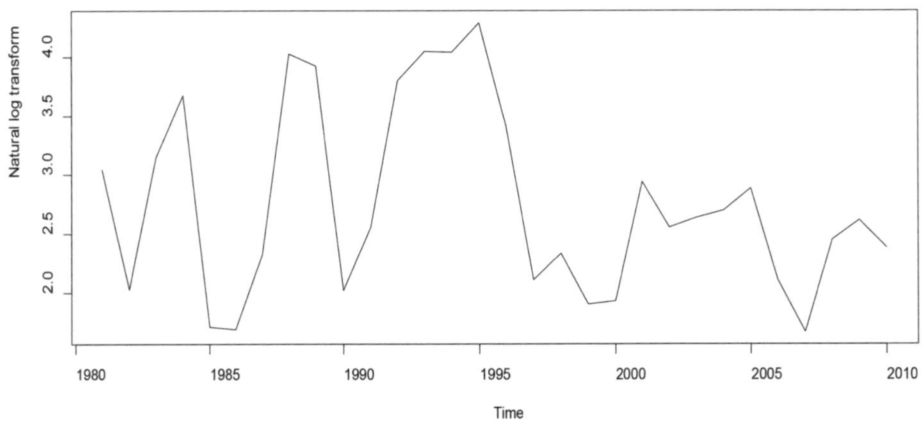

Figure 2.
Plot of Natural Log transform of Inflation Rate from 1981 to 2010.

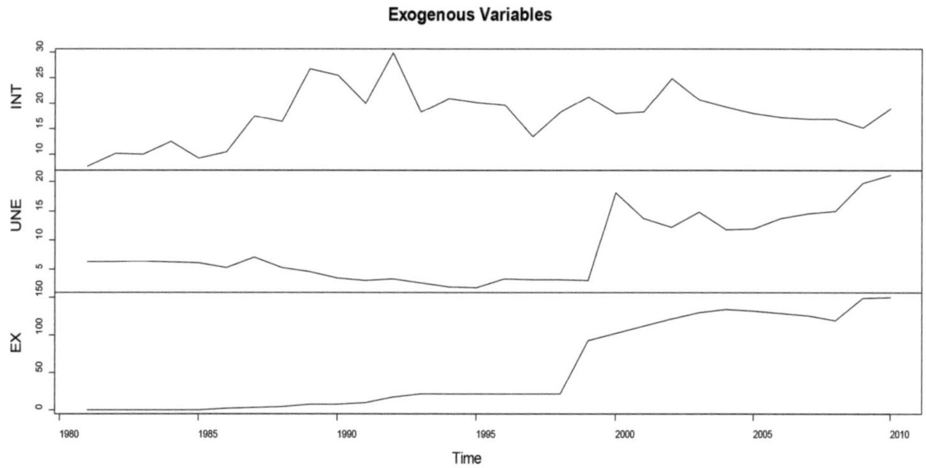

Figure 3.
The plots of interest rate (INT), unemployment rate (UNE) and exchange rate (EX) from 1981 to 2010.

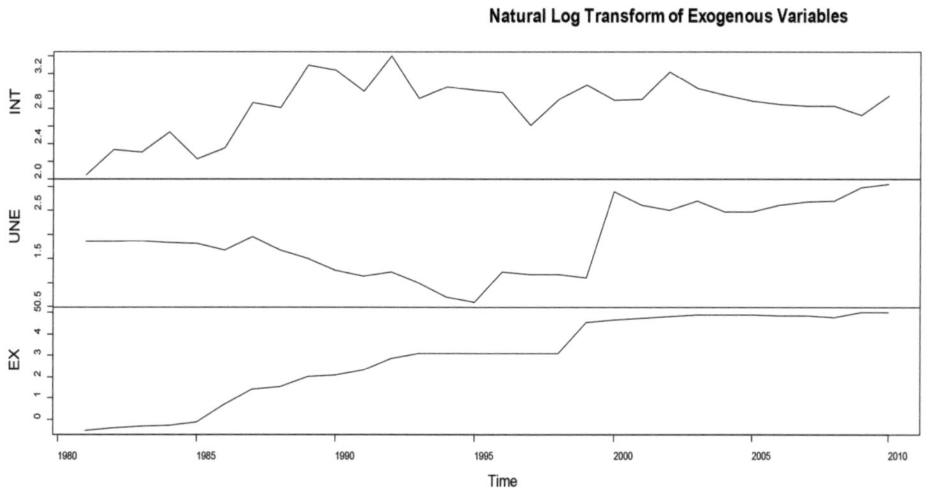

Figure 4.
The Plots of the natural transform of interest rate (INT), unemployment rate (UNE) and exchange rate (EX) from 1981 to 2010.

Figure 4 above presented the plots of the natural log transform of Interest Rate (INT), Unemployment Rate (UNE) and Exchange Rate (EX) from 1981 to 2010 in Nigeria. The log of interest rate shows some decrease 2002 to 2010 but for logs of unemployment rates and exchange rates shows an increase from 2002 to 2009. This situation about unemployment and exchange rates will definitely affect the standard of living in Nigeria if not properly control. This similar to **Figure 3** above.

Table 2 below presents the results of Jarque-Bera (JB) normality test of the inflation rate and the natural log transform of the inflation rate. The result revealed that inflation rate is not normally distributed since p-value = 0.02196 < 0.05. But the log transform of inflation rate is normally distributed since p-value = 0.3075 > 0.05. This test is necessary because the ARIMA and ARIMAX models are dependent on normal distribution.

	Inflation rate	Natural log transform of inflation rate
JB Test	7.6367	2.3586
P-Value	0.02196	0.3075

Table 2.
Normality test.

	Natural log transform of inflation rate at level	Natural log transform of inflation rate at 1st difference
ADF Test	−2.2624	−3.9483
P-Value	0.4719	0.02442
Remark	Not stationary	Stationary

Table 3.
ADF unit root test.

The **Table 3** above presents the unit root test using Augmented Dickey Fuller (ADF) test of the inflation rate. The ADF test is necessary in order to avoid spurious regression. The test revealed that the first difference of the log transform of inflation rate is stationary since p-value = 0.02442 < 0.05. This result imply that integration (I) must be added to the estimated ARIMA and ARIMAX.

Figures 5 and **6** presented the ACF and PACF of the log transform of inflation rate. Evidence revealed a combination of AR and MA processes for the inflation rate model.

The summary performances of the ARIMA and ARIMAX Model.

Table 4 below presents the in-sample performance of the ARIMA competing models. Among the ARIMA models, ARIMA(1,1,1) has the least values of RMSE and MAE. Hence ARIMA (1,1,1) outperformed the other ARIMA models while ARIMA (1,1,0) model is the worst. In addition, the coefficients of the ARIMA(1,1,0) are not significant (p-values>0.05) but the coefficients of ARIMA(1,1,0) and ARIMA(0,1,1) models are significant (p-values<0.05). The residual from the models are normally

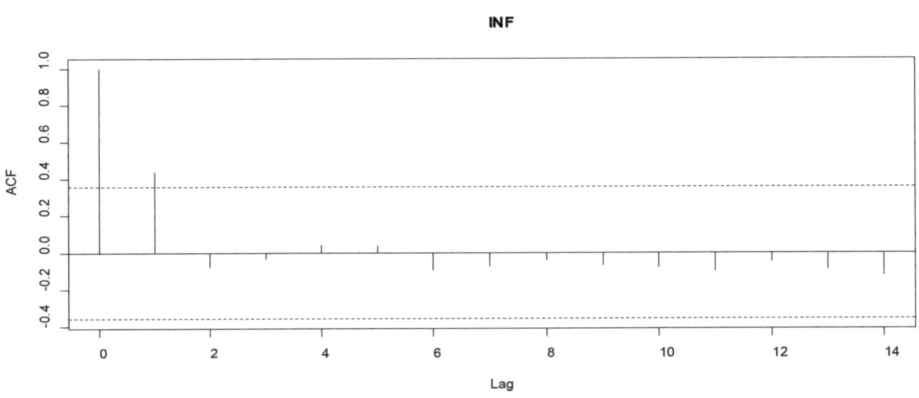

Figure 5.
ACF plot of the natural log transform of inflation rate.

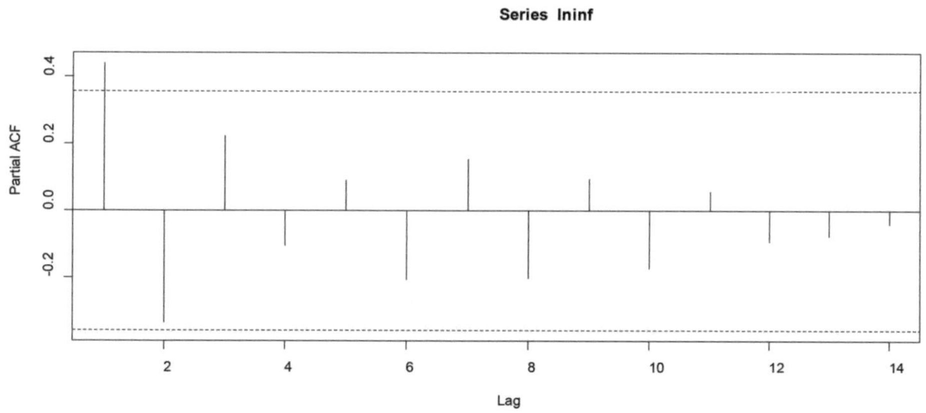

Figure 6.
PACF Plot of the natural log transform of inflation rate.

Model	RMSE	MAE	JB Test on Residual (P-values)	Adequacy test (Box-Ljung Test)
ARIMA(1,1,0)	0.8322	0.6125	0.6066	Adequate
ARIMA(1,1,1)	0.7200	0.5675	0.9961	Adequate
ARIMA(0,1,1)	0.8054	0.6617	0.6744	Adequate

Table 4.
In-sample performances of the ARIMA models.

distributed (p-values>0.05) while all the models passed the adequacy test (p-values>0.05).

Table 5 below presents the in-sample performance of the ARIMAX competing models. Among the ARIMA models, ARIMAX(1,1,1) and ARIMAX (0,1,1) has the least values of RMSE and MAE. Hence ARIMAX (1,1,1) and ARIMAX (0,1,1) models are preferred while ARIMAX (1,1,0) model is the worst. In addition, the coefficients of the ARIMAX(1,1,0) are not significant (p-values>0.05) while interest and exchange rates are positively related to inflation and unemployment rate is negatively related to inflation rate, though the coefficient of the exogenous variable are not significant (p-values>0.05). For ARIMAX (1,1,1) model the AR and MA coefficients are significant (p-values<0.05) while interest and exchange rates are positively related to inflation and unemployment rate is negatively related to inflation rate, though the coefficient of the exogenous variable are not significant (p-values>0.05). for ARIMAX(0,1,1) model, the MA coefficient is significant (p-value<0.05) while interest rate is positively related to inflation rate, but unemployment and exchange rates are negatively related to inflation rate, though the coefficient of the exogenous variable are not significant (p-values>0.05). The residual from the models are

Model	RMSE	MAE	JB Test on Residual (P-values)	Adequacy test (Box-Ljung Test)
ARIMAX(1,1,0)	0.8107	0.6401	0.8394	Adequate
ARIMAX(1,1,1)	0.7058	0.5519	0.961	Adequate
ARIMAX(0,1,1)	0.6810	0.5820	0.489	Adequate

Table 5.
In-sample performances of the ARIMAX models.

Model	RMSE	MAE
ARIMA(1,1,1)	0.3069	0.1970
ARIMAX(1,1,1)	0.3467	0.2777
ARIMAX(0,1,1)	0.3071	0.2336

Table 6.
Out-of-sample performances of the ARIMA and ARIMAX models.

Year	Actual Log Inflation Rate	ARIMA (1,1,1) Forecast of Log Inflation Rate	ARIMAX (1,1,1) Forecast of Log Inflation Rate	ARIMAX (0,1,1) Forecast of Log Inflation Rate
2011	2.332144	2.172615	2.112835	2.159863
2012	2.442347	2.236377	2.123062	2.053614
2013	2.140066	2.158553	2.014263	2.068257
2014	2.085672	2.154786	1.964497	1.998779
2015	2.104134	2.112283	1.994304	2.073538
2016	2.261763	2.090041	1.960178	2.012625
2017	2.803360	2.057202	2.056447	2.167942

Table 7.
Actual and forecast of log inflation rate from 2011 to 2017.

normally distributed (p-values>0.05) while all the models passed the adequacy test (p-values>0.05).

Table 6 above presented the out-of-sample forecast statistic of the preferred ARIMA and ARIMAX models from the in-sample forecast. The result revealed ARIMA (1,1,1) has the least values of RMSE and MAE. Hence ARIMA(1,1,1) model is preferred while ARIMAX (1,1,1) model is the worst model.

Table 7 above present the actual and forecast of log inflation rate from 2011 to 2011. The forecast revealed a fluctuations in inflation rates but evidence of reduction in inflation rates from 2012 to 2017 from the ARIMA(1,1,1). The plot of out-of-sample forecast from ARIMA(1,1,1), ARIMAX(1,1,1) and ARIMAX(0,1,1) are presented in **Figures** 7-9 respectively.

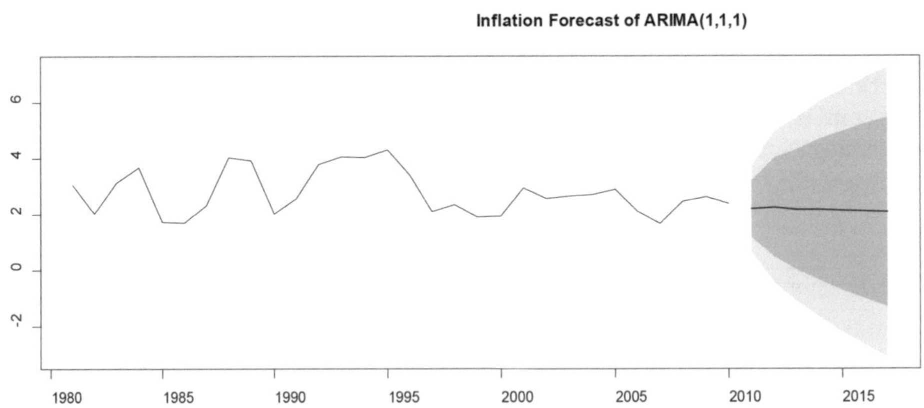

Figure 7.
Inflation forecast of ARIMA(1,1,1) from 2011 to 2017.

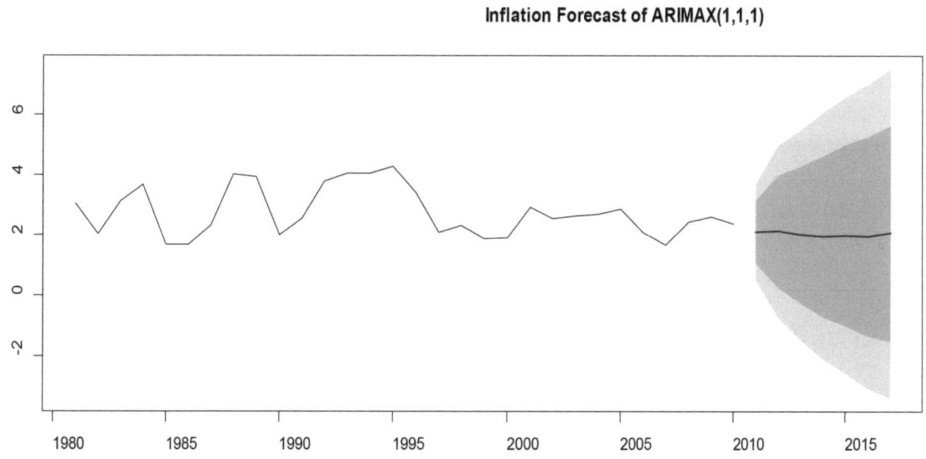

Figure 8.
Inflation forecast of ARIMAX(1,1,1) from 2011 to 2017.

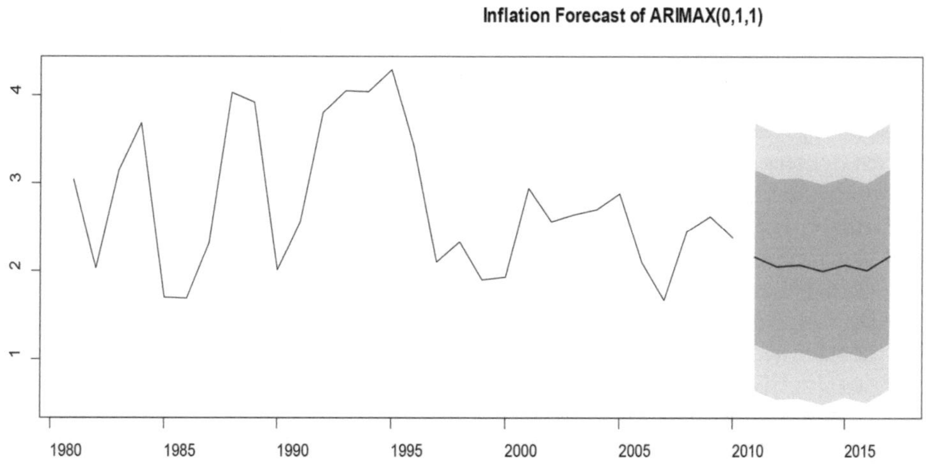

Figure 9.
Inflation Forecast of ARIMAX(0,1,1) from 2011 to 2017.

6. Conclusions

This book chapter concluded that:

ARIMAX(0,1,1) with RMSE of 0.6810 emerged as superior model for the in-sample forecast for forecasting inflation rate in Nigeria while ARIMA (1,1,1) emerged as a superior model for the out-of-sample forecast for inflation rate in Nigeria and its forecast for inflation revealed a negative growth in inflation in Nigeria. In addition, the entire models estimated are adequate and their residuals are normally distributed.

Based on the findings of this chapter, the following are recommended:

i. ARIMAX (0,1,1) model be used for in-sample forecast for inflation rate in Nigeria.

ii. ARIMA (1,1,1) model be used for out-of-sample forecast for inflation rate in Nigeria.

iii. Government should endeavor to formulate policy to reduce the negative effect of inflation rate on the citizenry and on the economy of Nigeria.

Acknowledgements

I wish to acknowledge my M.Sc. students that have worked under my supervision in the area of econometric time series analysis.

Conflict of interest

The Author declares no conflict of interest.

A. Appendix

R Codes

```
Endo<-ts(read.table("C:/Users/ADENOMON/Desktop/Inflation.txt",header=T),
start=c(1981,1),freq=1)
inftest<-ts(read.table("C:/Users/ADENOMON/Desktop/Inflatest.txt",
header=T),start=c(2011,1),freq=1)
Exoge<-ts(read.table("C:/Users/ADENOMON/Desktop/Exoge.txt",header=T),
start=c(1981,1),freq=1)
lnExoge<-log(Exoge)
adf.test(lnExoge[,1])
adf.test(lnExoge[,2])
adf.test(lnExoge[,3])
adf.test(diff(lnExoge[,1]))
adf.test(diff(lnExoge[,2]))
adf.test(diff(lnExoge[,3]))
fit1=Arima(lninf,order=c(1,1,0),include.constant=TRUE)
summary(fit1)
coeftest(fit1)
acf(fit1$residuals,main="ARIMA(1,1,0) Residuals")
jarqueberaTest(fit1$residuals)
Box.test(fit1$residuals,lag=12,type="Ljung-Box")
Box.test(fit1$residuals,lag=24,type="Ljung-Box")
qqnorm(fit1$residuals,main="Normal Q-Q Plot of ARIMA(1,1,0)")
qqline(fit1$residuals,main="Normal Q-Q Plot of ARIMA(1,1,0)")
fit2=Arima(lninf,order=c(1,1,1),include.constant=TRUE)
summary(fit2)
coeftest(fit2)
acf(fit2$residuals,main="ARIMA(1,1,1) Residuals")
jarqueberaTest(fit2$residuals)
Box.test(fit2$residuals,lag=12,type="Ljung-Box")
Box.test(fit2$residuals,lag=24,type="Ljung-Box")
qqnorm(fit2$residuals,main="Normal Q-Q Plot of ARIMA(1,1,1)")
qqline(fit2$residuals,main="Normal Q-Q Plot of ARIMA(1,1,1)")
fit3=Arima(lninf,order=c(0,1,1),include.constant=TRUE)
```

```
summary(fit3)
coeftest(fit3)
acf(fit3$residuals,main="ARIMA(0,1,1) Residuals")
jarqueberaTest(fit3$residuals)
Box.test(fit3$residuals,lag=12,type="Ljung-Box")
Box.test(fit3$residuals,lag=24,type="Ljung-Box")
qqnorm(fit3$residuals,main="Normal Q-Q Plot of ARIMA(0,1,1)")
qqline(fit3$residuals,main="Normal Q-Q Plot of ARIMA(0,1,1)")
fit4=Arima(lninf,order=c(1,1,0),xreg=lnExoge,include.constant=TRUE)
summary(fit4)
coeftest(fit4)
acf(fit4$residuals,main="ARIMAX(1,1,0) Residuals")
jarqueberaTest(fit4$residuals)
Box.test(fit4$residuals,lag=12,type="Ljung-Box")
Box.test(fit4$residuals,lag=24,type="Ljung-Box")
qqnorm(fit4$residuals,main="Normal Q-Q Plot of ARIMAX(1,1,0)")
qqline(fit4$residuals,main="Normal Q-Q Plot of ARIMAX(1,1,0)")
fit5=Arima(lninf,order=c(1,1,1),xreg=lnExoge,include.constant=TRUE)
summary(fit5)
coeftest(fit5)
acf(fit5$residuals,main="ARIMAX(1,1,1) Residuals")
jarqueberaTest(fit5$residuals)
Box.test(fit5$residuals,lag=12,type="Ljung-Box")
Box.test(fit5$residuals,lag=24,type="Ljung-Box")
qqnorm(fit5$residuals,main="Normal Q-Q Plot of ARIMAX(1,1,1)")
qqline(fit5$residuals,main="Normal Q-Q Plot of ARIMAX(1,1,1)")
fit6=Arima(lninf,order=c(0,1,1),xreg=lnExoge,include.constant=TRUE)
summary(fit6)
coeftest(fit6)
acf(fit6$residuals,main="ARIMAX(0,1,1) Residuals")
jarqueberaTest(fit6$residuals)
Box.test(fit6$residuals,lag=12,type="Ljung-Box")
Box.test(fit6$residuals,lag=24,type="Ljung-Box")
qqnorm(fit6$residuals,main="Normal Q-Q Plot of ARIMAX(0,1,1)")
qqline(fit6$residuals,main="Normal Q-Q Plot of ARIMAX(0,1,1)")
fit1.pred<-forecast(fit1,h=7)
accuracy(fit1.pred$pred,)
coeftest(fit3)
acf(fit3$residuals,main="ARIMA(0,1,1) Residuals")
jarqueberaTest(fit3$residuals)
Box.test(fit3$residuals,lag=12,type="Ljung-Box")
Box.test(fit3$residuals,lag=24,type="Ljung-Box")
qqnorm(fit3$residuals,main="Normal Q-Q Plot of ARIMA(0,1,1)")
qqline(fit3$residuals,main="Normal Q-Q Plot of ARIMA(0,1,1)")
fit3=Arima(Dlninf,order=c(0,1,1),include.constant=TRUE)
summary(fit3)
coeftest(fit3)
acf(fit3$residuals,main="ARIMA(0,1,1) Residuals")
jarqueberaTest(fit3$residuals)
Box.test(fit3$residuals,lag=12,type="Ljung-Box")
```

```
Box.test(fit3$residuals,lag=24,type="Ljung-Box")
qqnorm(fit3$residuals,main="Normal Q-Q Plot of ARIMA(0,1,1)")
qqline(fit3$residuals,main="Normal Q-Q Plot of ARIMA(0,1,1)")
Exogetest<-ts(read.table("C:/Users/ADENOMON/Desktop/Exogetest.txt",
header=T),start=c(2011,1),freq=1)
lnExotest<-log(Exogetest)
lninftest<-log(inftest)
accuracy(fit1.pred,lninftest)
fit2.pred<-forecast(fit2,h=7)
accuracy(fit2.pred,lninftest)
plot(fit2.pred, main="Inflation Forecast of ARIMA(1,1,1)")
fit5.pred<-forecast(fit5,h=7,xreg=lnExotest)
accuracy(fit5.pred,lninftest)
plot(fit5.pred, main="Inflation Forecast of ARIMAX(1,1,1)")
fit6.pred<-forecast(fit6,h=7,xreg=lnExotest)
accuracy(fit6.pred,lninftest)
plot(fit6.pred, main="Inflation Forecast of ARIMAX(0,1,1)")
```

Author details

Monday Osagie Adenomon[1,2,3,4*] and Felicia Oshuwalle Madu[5]

1 Department of Statistics, Nasarawa State University, Nigeria

2 NSUK-LISA Stat Lab, Nasarawa State University, Nigeria

3 Chair, International Association of Statistical Computing (IASC) African Members Group, Nigeria

4 Foundation of Laboratory for Econometrics and Applied Statistics of Nigeria (FOUND-LEAS-IN-NIGERIA)

5 National Bureau of Statistics, Abuja, Nigeria

*Address all correspondence to: adenomonmo@nsuk.edu.ng

References

[1] Kirchgässner G, Wolters J. Introduction to Modern Time Series Analysis. New York: Springer Books; 2017

[2] Adenomon MO. Modelling and forecasting unemployment rates in Nigeria using ARIMA model. FUW Trends in Science & Technology Journal. 2017;**2**(1B):525-531

[3] Cooray TMJA. Applied Time series Analysis and Forecasting. New Delhi: Narosa Publising House; 2008

[4] Box GEP, Jenkins GM. Time Series Analysis, Forecasting and Control. San Francisco: Holden-Day; 1970

[5] Kongcharoen C, Kruangpradit T. Autoregressive integrated moving average with explanatory variable (ARIMAX) model for Thailand export. A paper presented at the 33rd International Symposium on Forecasting, South Korea, June 2013. 2013

[6] Lipsey RG, Chrystal KA. Principles of Economics. 9th ed. United States: Oxford University Press; 1999. pp. 529-544

[7] Eitrheim Ø, Husebø TA, Nymoen R. Empirical comparison of inflation models' Fprecast accuracy. In: Clement MP, Hendry DF, editors. A Companion to Economic Forecasting. USA: Blackwell Publishing; 2004. pp. 354-385

[8] Omotor DG. Relationship between inflation and stock market returns. CBN Journal of Applied Statistics. 2010;**1**(1): 1-16

[9] Shittu OI, Yaya OS. On fractionally integrated logistic smooth transitions in time series. CBN Journal of Applied Statistics. 2011;**2**(1):1-13

[10] Abraham TW. Stock market reaction to selected macroeconomic variables in the Nigerian economy. CBN Journal of Applied Statistics. 2011;**2**(1):61-70

[11] Musa Y, Gulumbe SU. Analyzing Nigeria inflation and government revenues using ARDL approach. Nigerian Statistical Association 2014 Annual Conference Proceedings. 2014. pp. 195-209.

[12] Furuoka F. Unemployment and inflation in the Philippines. Philippine Journal of Development. 2008;**XXXV** (1):93-106

[13] Omar MRB, Sarkar HK. Relationship between commodity prices and exchange rate in light of global financial crisis: Evidence from Australia. International Journal of Trade, Economics and Finance. 2013;**4**(5): 265-269

[14] Mohaddes K, Raissi M. Does inflation slow long-run growth in India. IMF Working Paper. WP/14/222. 2014

[15] Mida J. Forecasting Exchange Rates: A VAR Analysis. Institute of Economic Studies. Charles University in Prague; 2013

[16] Nastansky A, Strohe HG. Public debt, money and consumer prices: A vector error correction model for Germany. Econometrics. 2015;**1**(47):9-31

[17] Gillitzer C. The Sticky Information Phillips Curve: Evidence for Australia. Reserve Bank of Autralia, RDP 2015-04. 2015

[18] Stock JH, Watson MW. Forecasting inflation. Journal of Monetary Economics. 1999;**44**:293-335

[19] Bougas C. Forecasting Air Passenger Traffic Flows in Canada: An Evaluation of Time Series Models and Combination Methods. Quebec, Canada: Universite Laval; 2013

[20] Adenomon MO, Tela MN. Development of simple time series model and ARIMA model for inflation rates in Nigeria. Journal of Natural and Applied Sciences. 2017;**5**(2):210-223

[21] Styrvold H, Nereng K. Forecasting Rental Rates for Norwegian Commercial real Estate. Norwegian Business School; 2011

[22] Amadeh H, Amini A, Effati F. ARIMA and ARFIMA prediction of Persian Gulf Gas-Oil F.O.B. Investment Knowledge. 2013;**2**(7):212-231

[23] Avuglar RK, Adu-Poku KA, Harris E. Application of ARIMA models to road traffic accident cases in Ghana. International Journal of Statistics and Applications. 2014;**4**(5):233-239

[24] Moshiri S, Foroutan F. Forecasting non linear crude oil future prices. The Energy Journal. 2006;**27**(4):81-96

[25] Engle RF, Granger CWJ. Co-integration and error correction: representation, estimation and testing. Econometrica. 1987;**55**(2):251-276

[26] Ajayi A, Mougoue M. On the dynamic relationship between stock prices and exchange rates. Journal of Financial Research. 1996;**9**(2):193-207

[27] Salvatore D, Reagle D. Schaum's Outline of Theory and Problems of Statistics and Econometrics. 2nd ed. New York: McGraw-Hill; 2002

[28] Dobre I, Alexandru AA. Modelling unemployment rate using Box-Jenkins procedure. Journal of Applied Quantitative Methods. 2008;**3**(2):156-166

[29] Gujarati DN. Basic Econometrics. 4th ed. New Delhi: The McGraw-Hill Co; 2003. pp. 835-848

[30] Ljung GM, Box GEP. On a measure of lack of fit in time series models. Biometrika. 1978;**69**(297):303

[31] Caraiani P. Forecasting Romanian GDP using a BVAR model. Romanian Journal of Economic Forecasting. 2010; **4**:76-87

[32] Central Bank of Nigeria (CBN). 2017 CBN Statistical Bulletin. Nigeria; 2017

Section 8

Diffusion Processes

Chapter 8

The $\mathbb{L}_2-$ Structure of Subordinated Solution of Continuous-Time Bilinear Time Series

Abdelouahab Bibi

Abstract

The models of stochastic subordination, or random time indexing, has been recently applied to model financial returns $(X(t))_{t\geq 0}$ exhibiting some characteristic periods of constant values for instance exchange rate. In reality, sharp and large variations for $X(t)$ do occur. These sharp and large variations are linked to information arrivals and/or represent sudden events and hence we have a model with jumps. For this purpose, by substituting the usual deterministic time t as a subordinator $(T(t))_{t\geq 0}$ in a stochastic process $(X(t))_{t\geq 0}$ we obtain a new process $(X(T(t)))_{t\geq 0}$ whose stochastic time is dominated by the subordinator $(T(t))_{t\geq 0}$. Therefore we propose in this paper an alternative approach based on a combination of the continuous-time bilinear (*COBL*) process subordinated by a Poisson process (that it is a Levy process) which permits us to introduce further randomness for the phenomena which exhibit either a speeded up or slowed down behavior. So, the main probabilistic properties of such models are studied and the explicit expression of the higher-order moments properties are given. Moreover, moments method (*MM*) is proposed as an estimation issue of the unknown parameters. Simulation studies confirm the theoretical findings and show that the *MM* method proposal can effectively reduce both the bias and the mean square error of parameter estimates.

Keywords: diffusion processes, subordination, Poisson process

1. Introduction

The non-linear time-continuous models were initially discussed by Mohler [1] in control theory and then rapidly extended to a time-series analysis by several authors (see [2] for review). One of the classes of non-linear time-continuous models which has attracted considerable attention of the researchers is the classes of bilinear diffusion processes which have been widely studied and considered in time series analysis and in the theory of stochastic differential equations (*SDE*). For instance, among others, Le Breton and Musiela [3] and Bibi and Merahi [4] have considered a process $(X(t))_{t\geq 0}$ generated by the following *SDE*

$$dX(t) = (\alpha X(t) + \mu)dt + (\gamma X(t) + \beta)dw(t), t \geq 0, X(0) = X_0 \qquad (1)$$
$$= \mu(X(t))dt + \sigma(X(t))dw(t)$$

denoted hereafter by *COBL* (1) in which $\mu(x) = \alpha x + \mu$ and $\sigma(x) = \gamma x + \beta$ are respectively the drift and diffusion functions representing respectively the conditional mean and variance of the infinitesimal change of $X(t)$ at time $t.(w(t))_{t\geq 0}$ is a real standard Brownian motion defined on some basic filtered space $(\Omega, \mathcal{A}, (\mathcal{A}_t)_{t\geq 0}, P)$ and $E\{X(t)dw(t)\} = 0$. The initial condition $X(0)$ of $X(t)$ can be either deterministic or random variable defined on $(\Omega, \mathcal{A}, P)$ independent of w such that $E\{X(0)\} = m_1(0)$ and $Var\{X(0)\} = K_X(0)$. However, the distribution of stochastic processes $X(t)$ solution of (1) evaluated at random times process say $T(t)$, are receiving increasing attention in various applied fields. Some examples we have in mind are:

1. in reliability theory, the life span of some items subjected to certain accelerated conditions,

2. in econometrics, the composition of the prices at short intervals on a speculative market,

3. in queuing theory, the number of customers arriving at random times to some facility where they receive service of some kind and then depart,

4. in statistics, for the random sampling of stochastic processes.

One of the first papers in this field is by Lee and Whitmore [5], who studied general properties of processes delayed by randomized times. In the literature, due to the interesting properties of the Poisson process, its popularity, and applicability, various researchers have generalized it in several directions; e.g., compound Poisson processes and/or weighted Poisson distributions, special attention is given to the case of a Poisson process with randomized time or Poisson subordinator, i.e.; the time process is supposed to be a subordinator – Poisson process with nondecreasing sample paths. Most published research involving this approach, Clark [6] and German and Ane [7].

In this paper, our interest lies in the statistical inference of the parameters involved in the diffusion process defined in (1) and in its subordination by a Poisson process. Diffusion processes estimation has been widely studied in the statistical literature by many authors under several restrictions (see [8] for a survey). The major approach used in parameters estimation is the maximum likelihood method which in general presents a difficulty to obtaining a tractable expression for the transition densities. So, certain econometric methods have been recently proposed. Hence, parameters estimation of continuous-time processes can be achieved through the use for instance the moments method (*MM*) and/or its generalization (*GMM*). These methods are useful for modeling some events that occur randomly over a the fixed period of time or in a fixed space chaotic subordination by assuming a Poisson process for the subordinating variable for *COBL*(1,1) and hence some statistical and probabilistic properties are studied. For this purpose, in next section we describe some theoretical framework for certain specification of *COBL*(1,1). More precisely, we discuss the condition of their existence, uniqueness, and their distribution. The moments properties of *COBL*(1,1) process are presented in section 3 followed by its extended to that subordination by a Poisson process. In section 4 we discuss the properties of the subordinated process, in particular, its moment properties and its distribution of the subordinated version. An estimation issue based on *MM* and on *GMM* (considered as a benchmark) are presented in section 5, substantially enriched by the asymptotic properties of such estimations. In section 6, Monte-Carlo simulation is carried out through a simulation study of $COBL(1,1)$ and its subordinated process. The end section is for the conclusions.

2. Theoretical background

The *SDE* (1,1) covers many models commonly used in the literature. Some specific examples among others are:

1. $COGARCH(1,1)$: This class of processes is defined as a *SDE* by $dX(t) = \sigma(t)dB_1(t)$ with $d\sigma^2(t) = (\mu - \alpha\sigma^2(t))dt + \gamma\sigma^2(t)dB_2(t), t > 0$ where B_1 and B_2 are independent Brownian motions, $\mu > 0$, $\alpha \geq 0$, and $\gamma \geq 0$. So, the stochastic volatility equation can be regarded as a particular case of (1) by assuming $\beta = 0$. (see [9]).

2. $CAR(1)$: This classes of *SDE* may be obtained by assuming $\gamma = 0$ (see [10]).

3. Gaussian Ornstein-Uhlenbeck (*OU*) process: The *OU* process is defined as

$$dX(t) = (\mu + \alpha X(t))dt + \beta dw(t), t \geq 0 \tag{2}$$

with the diffusion parameter $\beta > 0$. So it can be obtained from (1) by assuming $\gamma = 0$ (see [10] and the reference therein).

4. Geometric Brownian motion (*GBM*): This class of processes is defined as a $\mathbb{R}-$ valued solution process $(X(t))_{t \geq 0}$ of $dX(t) = \alpha X(t)dt + \gamma X(t)dw(t), t \geq 0$. So it can be obtained from (1) by assuming $\beta = \mu = 0$ (see [11]).

2.1 Existence of ergotic and stationary solutions

The existence of solution process of equation (1), was investigated by several authors, for instance, Iglói and Terdik [12] have studied the same model driven by fractional Brownian innovation. A class of *COBL* with time-varying coefficients was studied by Le Breton and Musiela [3], Bibi and Merahi [4] and Leon and Perez-Abreu [13]. Moreover, there are several monographic which discuss the theoretical probabilistic and statistical properties (interested readers are advised to see [14, 15] and the references therein). Hence, a Markovian Itô solution of *SDE* (1) is given by

$$X(t) = \Phi(t)\left\{X(0) + (\mu - \gamma\beta)\int_0^t \Phi^{-1}(s)ds + \beta\int_0^t \Phi^{-1}(s)dw(s)\right\}, a.s., \tag{3}$$

where $\Phi(t) = \exp\left\{\left(\alpha - \frac{1}{2}\gamma^2\right)t + \gamma w(t)\right\}$ is the fundamental process solution (see e.g., [14] chapter 8) and its first and second moments functions $\Psi(t) = E\{\Phi(t)\} = \exp\{\alpha t\}$ and $\phi(t) = E\{\Phi^2(t)\} = \exp\{(2\alpha + \gamma^2)t\}$. The key tool in studying the asymptotic stability of solution (3) is the top-Lyapunov exponent defined by $\lambda_L = \lim\sup_{t\to+\infty} \frac{1}{t}\log|X(t)|$, so if it exists then λ_L controls the long-time asymptotic behavior of X. Indeed if $\lambda_L < +\infty$, *a.s* then for sufficiently large t, there exists a positive random variable ξ such that $|X(t)| \le \xi e^{\lambda_L t}$ and hence if $\lambda_L < 0$, then $\lim_{t\to+\infty} X(t) = 0$, *a.s.* Though the condition $\lambda_L < 0$ could be used as a sufficient condition for asymptotic stability, it is of little use for the practice of checking for stationarity of the solution (3). On the other hand, and in statistical applications, we often suggest conditions ensuring the existence of some moments for the process solution. This suggestion cannot be achieved by the top-Lyapunov exponent criterion. However, since the functions $\mu(x)$ and $\sigma(x)$ are locally Lipschitz, then the existence and uniqueness of stationary and ergodic solution process $(X(t))_{t\ge0}$ given by (3) is ensured by the integrability on $\mathbb{R}^+$ of the speed density $g(y) = \frac{1}{\sigma^2(y)}\exp\left\{2\int_1^y \frac{\mu(x)}{\sigma^2(x)}dx\right\}$ (see [16]) and that the density function $f(.)$ of the stationary distribution of a diffusion process (1) is proportional to $g(y)$. Moreover, the unique invariant probability is absolutely continuous with respect to the Lebesgue measure with a density function equal to g (up to a constant). Hence, the integrability on $\mathbb{R}^+$ of the function g may be discussed case by case in the following cases

1. $\gamma = 0$ and $\beta \neq 0$ (*OU* case), in this case $g(y) = C\exp\left\{\frac{\alpha}{\beta^2}\left(y + \frac{\mu}{\alpha}\right)^2\right\}$ for some positive constant C, and hence $g(y)$ is integrable on $\mathbb{R}^+$ if and only if $\alpha < 0$ for all $\mu \in \mathbb{R}$. Therefore we recognize a $\mathcal{N}\left(-\frac{\mu}{\alpha}, -\frac{\beta^2}{2\alpha}\right)$ for the invariant distribution of *OU* process and

$$f(y) = \frac{1}{\sqrt{-2\pi\frac{\beta^2}{2\alpha}}}\exp\left\{\frac{1}{\frac{\beta^2}{\alpha}}\left(y + \frac{\mu}{\alpha}\right)^2\right\}$$

2. $\beta = 0$, $\mu = 0$ (*GBM* case) in this case $g(y) = Cy^{2(\alpha-\gamma^2)/\gamma^2}$ and hence $g(y)$ is not integrable on $\mathbb{R}^+$, therefore there is no stationary and ergodic solution for *GBM* process.

3. $\beta = 0$, $\mu \neq 0$ (*COBL*(1,1) case) the function $g(y) = C\left(\frac{1}{y}\right)^{(\gamma^2-2\alpha)/\gamma^2+1}\exp\left\{-2\frac{\mu}{\gamma^2 y}\right\}$, the integrability conditions hold if and only if $\mu > 0$, and hence the unique ergodic and stationary solution exists on $\mathbb{R}^+$. Therefore, we recognize a inverse-gamma distribution noted $\mathcal{IG}(\delta, \theta)$ for the invariant distribution of *COBL*(1,1) process where the shape parameter $\delta = (\gamma^2 - 2\alpha)/\gamma^2 > 0$ and the scale parameter $\theta = 2\frac{\mu}{\gamma^2} > 0$ and

$$f(y) = \frac{\theta^\delta}{\Gamma(\delta)}y^{-\delta-1}\exp\{-\theta/y\};\ y > 0.$$

The inverse-gamma distribution appears in Bayesian inference, in a natural way, as the posterior distribution of the variance in normal sampling. The process associated with this parametrization is often referred to *GARCH* diffusion models. Note that the $\mathcal{IG}$ distribution nests some well-known distributions such as the Inverse Exponential, Inverse χ^2 and Scaled Inverse χ^2 distributions.

In view of the above discussion, and since we are interested in the stationary non-Gaussian solution of (1), therefore it is necessary to assume throughout the rest of the paper that the parameters, α, μ, γ and β are subject to the following assumption:

Assumption 1. $\alpha\beta \neq \gamma\mu$, $\mu > 0$, $\gamma \neq 0$ *and* $2\alpha + \gamma^2 < 0$.

Remark 2.1. *The case* $\beta \neq 0$ *may be treated as that* $\beta = 0$ *by considering the affine transformation* $\tilde{X}(t) = \frac{\mu}{\gamma\mu-\alpha\beta}(\gamma X(t) + \beta)$. *On the contrary, the condition* $\gamma\mu \neq \alpha\beta$ *must be hold, otherwise the equation* (1) *has only a degenerate solution, i.e.,* $X(t) = -\frac{\beta}{\gamma} = -\frac{\mu}{\alpha}$. *The solution* (3) *is however Markovian when* $\beta \neq 0$, *otherwise the solution process is neither a standarized diffusion process nor a martingale. In contrast, if* $\gamma = 0$ *(OU process), the stochastic term is a martingale and hence it has a vanishing expectation. So, In the sequel, and without loss of generality we shall assume, that* $\beta = 0$, *i.e.,*

$$dX(t) = (\alpha X(t) + \mu)dt + \gamma X(t)dw(t), t \geq 0, X(0) = X_0, \tag{4}$$

and this equation will be the subject of our investigation so it is noted hereafter $COBL(1,1)$.

Remark 2.2. *In OU diffusion with* $\mu = 0$, *its solution is given by* $X(t) = X(0)e^{-\alpha t} + \beta \int_0^t e^{-\alpha(t-s)}dw(s)$, $t \geq 0$ *and its invariant probability distribution is Gaussian with mean 0 and variance* $\frac{\gamma^2}{2\alpha}$. *Moreover under the Assumption 1,*

1. *If* $X(0$ *is real constant, we have* $E\{X(t)\} = X(0)e^{-\alpha t}$, $Cov(X(t), X(t+h)) = \frac{\beta^2}{2\alpha}\left(e^{-\alpha h} - e^{-\alpha(2t+h)}\right)$, *and as* $t \to +\infty$, $E\{X(t)\} = 0$ *and* $Cov(X(t), X(t+h)) = \frac{\gamma^2}{2\alpha}e^{-\alpha h}$, $h \geq 0$.

2. *If* $X(0)$ *is random variable, then* $E\{X(t)\} = E\{X(0)\}e^{-\alpha t}$, $Cov(X(t), X(t+h)) = e^{-(2\alpha t+h)}Var\{X(0)\} + \frac{\beta^2}{2\alpha}\left(e^{-\alpha h} - e^{-\alpha(2t+h)}\right)$, *and as* $t \to +\infty$, $E\{X(t)\} = 0$, *and* $Cov(X(t), X(t+h)) = \frac{\gamma^2}{2\alpha}e^{-\alpha h}$, $h \geq 0$.

Remark 2.3. *In GBM with* $X(0) > 0$, *its solution is given by* $X(t) = \exp\left\{(\alpha - \frac{1}{2}\gamma^2)t + \gamma w(t)\right\}X(0)$. *So, the distribution of* $X(t)$ *given* $X(0)$ *is* log *normal with* $E\{X(t)\} = E\{X(0)\}e^{\alpha t}$ *and* $Var\{X(t)\} = E\{X^2(0)\}e^{2\alpha t}\left\{e^{\gamma^2 t} - 1\right\}$. *Hence, for any* $k \in \mathbb{R}$, *we have* $E\{X^k(t)\} = E\{X^k(0)\}\exp\left\{k\left(\alpha - \frac{\gamma^2}{2}\right)t + k^2\frac{\gamma^2}{2}t\right\}$, *so* $E\{X^k(t)\} \to +\infty$ *as* $t \to \infty$ *whenever* $\left(\alpha - \frac{\gamma^2}{2}\right)k + \frac{\gamma^2}{2}k^2 > 0$. *Additionally,*

a. If $\alpha > \frac{1}{2}\gamma^2$, then $\lim_{t\to+\infty} X(t) = +\infty$,

b. If $\alpha < \frac{1}{2}\gamma^2$, then $\lim_{t\to+\infty} X(t) = 0$,

c. If $\alpha = \frac{1}{2}\gamma^2$, then asymptotically $(X(t))_{t\geq 0}$ switches arbitrary between large and small positive values.

3. Moments properties of $COBL(1,1)$ process

In the sequel, we shall focus on the popular sub–model (4). The popularity of such a model comes from its solution in terms of stochastic integral, i.e.,

$$X(t) = X(0)\Phi(t) + \mu\int_0^t \Phi(t)\Phi^{-1}(s)d(s), t \geq 0. \tag{5}$$

or equivalently

$$X(t) = X(0) + \int_0^t (\alpha X(s) + \mu)ds + \gamma\int_0^t X(s)dw(s), t \geq 0, X(0) = X_0 \tag{6}$$

It is easy verified that the process $(X(t))_{t \geq 0}$ as defined by (5) satisfies (1) for any α, μ, γ, $\beta = 0$ and $X(0)$, it is the unique strong solution to (1). The following proposition summarizes the second-order properties.

Proposition 3.1. *If $X(0)$ is a random variable, then under the Assumption 1, we have*

1. $m(t) = E\{X(t)\} = \Psi(t)\left(E\{X(0)\} + \mu\int_0^t \Psi^{-1}(s)ds\right)$ and as $t \to +\infty$, $E\{X(t)\} = m = -\frac{\mu}{\alpha} > 0$.

2. For any $h \geq 0$, $K(t, t+h) = Cov(X(t), X(t+h)) = \Psi(h)K(t)$, where $K(t) = K(t,t)$ is the variance function given by $K(t) = \phi(t)\left\{K(0) + \gamma^2\int_0^t \phi^{-1}(s)m^2(s)ds\right\}$, so as $t \to +\infty$, $K(t) = -m^2\frac{\gamma^2}{2\alpha+\gamma^2}$. Hence $K(t, t+h) = -\Psi(h)m^2\frac{\gamma^2}{2\alpha+\gamma^2}$ and the correlation function is however $\rho(h) = e^{\alpha h}$. Therefore asymptotic stationary $COBL(1,1)$ process has autocorrelation function similar to a $CAR(1)$ processes.

Proof.

1. The first formula follows directly from (5).

2. The derivation of the second formula may be derived upon the observation that $X(t) - m(t) = \Psi(t)Y(t)$ where $dY(t) = \{\gamma Y(t) + \gamma\Psi^{-1}(t)m(t)\}dw(t)$ or equivalently $Y(t) = Y(0) + \int_0^y \{\gamma Y(t) + \gamma\Psi^{-1}(t)m(t)\}dw(t)$ with $Y(0) = X(0) - m(0)$ (see Bibi and Merahi for further details). So for any $h \geq 0$, we have $E\{Y(t)Y(t+h)\} = K(0) + \int_0^t \{\gamma^2 E\{Y^2(u)\} + \gamma^2\Psi^{-2}(u)m^2(u)\}du = E\{Y^2(t)\}$. Moreover, taking $h = 0$ and noting that $\Psi^2(t)E\{Y^2(t)\} = K(t)$ and $\Psi(t)\Psi(t+h)E\{Y(t)Y(t+h)\} = K(t, t+h)$, we obtain $K(t, t+h) = \Psi(h)K(t)$ and $K(t) = \Psi^2(t)Var\{X(0)\} + \Psi^2(t)\int_0^t \Psi^{-2}(u)\{\gamma^2 K(u) + \gamma^2 m^2(u)\}du$. Since $K(t) = \Psi^2(t)E\{Y^2(t)\}$, then $dK(t) = 2\Psi^2(t)E\{Y^2(t)\}dt + \Psi^2(t)dE\{Y^2(t)\}$ where

$dE\{Y^2(t)\} = \gamma^2 E\{Y^2(t)\} + \gamma^2 \Psi^{-2}(t) m^2(t)$. Thus $dK(t) = (2\alpha + \gamma^2)K(t) + \gamma^2 m^2(t)$. By solving the last differential equation, the expression of the variance follows. The rest of the proof fellows immediately by the dominated convergence Theorem.

Remark 3.2. *If $X(0)$ is a real constant, then the mean and variance of $(X(t))_{t \geq 0}$ reduces to*

$$m(t) = E\{X(t)\} = \Psi(t)\left(X(0) + \mu \int_0^t \Psi^{-1}(s)ds\right) \text{and} Var\{X(t)\} = \gamma^2 \phi(t) \int_0^t \phi^{-1}(u) m^2(u) du.$$

Moreover the $K(t, t+h)$ depends in general on time and on initial condition, thus the $COBL(1,1)$ process is not stationary but is asymptotically stationary. Except, for instance, in the following cases:

1. *Apart from assumption 1, if $E\{X(0)\} = -\frac{\mu}{\alpha}$, $\gamma^2 = -2\alpha$ and every $K(0)$ then the process $(X(t))_{t \geq 0}$ is second-order stationary.*

2. *If $E\{X(0)\} = -\frac{\mu}{\alpha}$ and $K(0) = \frac{(\gamma\mu)^2}{\alpha^2(2\alpha+\gamma^2)}$ then the process $(X(t))_{t \geq 0}$ is second-order stationary.*

3.1 Higher-order moment of $COBL(1,1)$ process

In what follows, we consider the function $f(x) = x^n$, then $f(X(t))$ is also an Itô's process. Applying Itô's formula on $f(X(t))$, we have

$$\begin{aligned} df(X(t)) &= f'(X(t))dX(t) + \frac{1}{2} f''(X(t))(dX(t))^2 \\ &= f'(X(t))\mu(X(t))dt + f'(X(t))\sigma(X(t))dw(t) + \frac{1}{2} f''(X(t))\sigma^2(X(t))d(t) \end{aligned}$$

which results to $dX^n(t) = \left(a_n X^n(t) + b_n X^{n-1}(t)\right)dt + c_n X^n(t) dw(t)$ or equivalently

$$X^n(t) = X^n(0) + \int_0^t \left(a_n X^n(s) + b_n X^{n-1}(s)\right) ds + c_n \int_0^t X^n(s) dw(s) \tag{7}$$

where $a_n = n\alpha + \frac{n(n-1)}{2}\gamma^2$, $b_n = n\mu$ and $c_n = n\gamma$. Due to stationarity and the fact that the last term of equation (7) is a zero mean martingale, then the moments of invariant distribution satisfy

$$E\{X^n(t)\} = -a_n^{-1} b_n E\{X^{n-1}(t)\} = (-1)^n \prod_{i=1}^n a_i^{-1} b_i. \tag{8}$$

The above equation allows us to find the moments of the invariant probability distribution for the Markov process generated by (5) for example $E\{X(t)\} = -\frac{\mu}{\alpha}$, $E\{X^2(t)\} = \frac{2\mu^2}{\alpha(2\alpha+\gamma^2)}$ and $Var(X(t)) = -\frac{(\mu\gamma)^2}{\alpha^2(2\alpha+\gamma^2)}$.

Example 3.1. *As already pointed out in the above section, the unique invariant probability distribution for the stationary solution of* (5) *has the form* $\{signG^{-1} - \frac{\mu}{\alpha}\}$ *where G has Gamma-distribution* $G(a, b)$ *with* $a = (\gamma^2 - 2\alpha)/\gamma^2$ *the shape parameter,* $b = \frac{\gamma^2}{2\mu}$ *is the scale parameter and the density* $f(x) = \frac{1}{\Gamma(a)b^a}x^{a-1}\exp\{-x/b\}$, $x > 0$. *So simple computation give* $E\{G\} = ab$ *and* $Var\{G\} = ab^2$ *however* $E\{\mathcal{G}^{-1}\} = -\frac{\mu}{\alpha}$ *and* $Var(\mathcal{G}^{-1}) = -\frac{\mu^2\gamma^2}{\alpha^2(2\alpha+\gamma^2)}$. *More generally for* $a > n$ *we have* $E\{\mathcal{G}^{-n}\} = \left(\frac{2\mu}{\gamma^2}\right)^n \prod_{i=1}^n (a-i)^{-1}$. *However, the above expression coincides with* (8).

Now, define $m_n(t,x) = E\{X^n(t)|X(0) = x\}$ to represent the n-*th* conditional moment of the process $(X(t))_{t\geq 0}$ defined by (5) for $n = 0, 1, 2, \ldots$ with $m_0(t,x) = 1$. Then simple manipulation of conditional expectation shows that $m_n(t,x)$ satisfy the following first-order recursive differential equation

$$dm_n(t,x) = a_n m_n(t,x)dt + b_n m_{n-1}(t,x)dt, \tag{9}$$

its solution is given in the following proposition

Proposition 3.3. *Suppose that the constants* $a_0, a_1, a_2, \ldots a_n$ *are distinct. Then under the Assumption 1, the solution of* (9) *for* $n = 0, 1, 2, \ldots$ *is given by* $m_n(t,x) = \sum_{i=0}^n \xi_i(n)e^{a_i t}$ *where* $\xi_i(n)$ *satisfies the recursion*

$$\xi_i(n) = \sum_{j=0}^{i} B_{j+1}^{(n)} A_{i,j}^{(n)} x^j, B_{j+1}^{(n)} = \prod_{k=j+1}^{n} b_k, A_{i,j}^{(n)} = \prod_{\substack{k=j \\ j \neq i}}^{n} \frac{1}{a_i - a_k} \tag{10}$$

with the convenient $B_{n+1}^{(n)} = 1$, $A_{n,n}^{(n)} = 1$.

Proof. See Bibi and Merahi [17].

Example 3.2. *The first and the second conditional moments are*

$$m_1(t,x) = -\frac{b_1}{a_1} + P_0(x)e^{a_1 t} \text{and} m_2(t,x) = \frac{b_1 b_2}{a_1 a_2} + P_1(x)e^{a_1 t} + P_2(x)e^{a_2 t}$$

where $P_0(x) = \left(\frac{b_1}{a_1} + x\right)$, $P_1(x) = \left(\frac{b_1 b_2}{a_1(a_1 - a_2)} + \frac{b_2}{(a_1 - a_2)}x\right)$ *and* $P_2(x) = \left(\frac{b_1 b_2}{a_2(a_2 - a_1)} + \frac{b_2}{(a_2 - a_1)}x + x^2\right)$.

Remark 3.4. *Note that when* $\alpha + \frac{n-1}{2}\gamma^2 < 0$ *for any n, the* $m_n(t,x)$ *converges as* $t \to +\infty$ *to unconditional moments* $E\{\mathcal{IG}^n\}$. *Moreover, when* $(X(t))_{t\geq 0}$ *is a GBM process* $m_n(t,x)$ *reduces to* $m_n(t,x) = x^n e^{a_n t}$ *because polynomial* $B_{j+1}^{(n)} = 0$ *for any* $j < n$ *and* $B_{n+1}^{(n)} = 1$. *Additionally, since for any* $n \geq 1$, $m_n(t,x)$ *depends on time, thus COBL*(1,1) *process with initial condition is non stationary, however it is asymptotically stationary.*

4. Subordinated COBL(1,1) process

The main idea of subordination (or change of time method) is to find a simple representation for $(X(t))_{t\geq 0}$ with a complicated structure, using some simple process and subordinator process $(T(t))_{t\geq 0}$. For example, if we consider a Brownian motion

$(w(t))_{t\geq 0}$ as a simple process and $(X(t))_{t\geq 0}$ that satisfies the stochastic differential equation (4) as a complicated process, then the question is: can we represent $(X(t))_{t\geq 0}$ in the following form $X(t) = w(T(t))$? In many cases, the answer is "yes" (see [18]). Hence, in this paper, we propose that T is represented by a homogeneous Poisson process.

4.1 Poisson counting process

The Poisson counting process, $\{N(t); t > 0\}$ consists of a nonnegative integer random variable $N(t)$ and satisfy the following definition

Definition 4.1. *A Poisson process $(N(t))_{t>0}$ is a counting process with the following additional properties*

1. $N(0) = 0$

2. *The process has stationary and independent increments.*

3. $P(N(t) = n) = \frac{(\lambda t)^n}{n!} \exp(-\lambda t)$ for $t > 0$ *and* $n = 0, 1, ..$ *the parameter λ is called the rate of the Poisson process.*

Remark 4.1. *Note that $N(t)$ is not a martingale but $N(t) - \lambda t$ it is. Moreover, in general, the "intensity" quantity λt may be replaced by a function $\lambda(t)$ which may be stochastic, to obtain an inhomogeneous Poisson process. It is worth noting that the definition 4.1 is quite close to the definition of the Wiener process and therefore have a similar method of approaching the simulation.*

Recalling that the probability generating function of $(N(t))_{t>0}$ is given by $E\{z^{N(t)}\} = \sum_{n=0}^{\infty} z^n P(N(t) = n) = e^{-\lambda t(1-z)}$. So by differentiation, we obtain the $4th-$

order non centered moments $v_k(t) = E\{N^k(t)\}, k = 1, \dots, 4$

$$v_1(t) = \lambda t, v_2(t) = (\lambda t)^2 + \lambda t, v_3(t) = (\lambda t)^3 + (\lambda t)^2 + \lambda t, v_4(t) = (\lambda t)^4 + 6(\lambda t)^3 + 7(\lambda t)^2 + \lambda t.$$

Moreover, the first four central moments $\mu_k(t) = E\left\{(N(t) - v_1(t))^k\right\}, k = 1, \dots, 4,$ are given by $\mu_1(t) = 0$, $\mu_2(t) = \lambda t$, $\mu_3(t) = \lambda t$ and $\mu_4(t) = 3(\lambda t)^2 + \lambda t$. Additionally, the skewness $Sk(t)$ and the excess kurtosis $Ku(t)$ coefficients of $N(t)$ are given by $Sk(t) = \frac{\mu_3^2(t)}{\mu_2^3(t)} = \frac{1}{\lambda t}$ and $Ku(t) = \frac{\mu_4(t)}{\mu_2^2(t)} = 3 + \frac{1}{\lambda t}$. Therefore the Poisson process is always a skewed and leptokurtic distribution for any $t > 0$.

4.2 Subordinated COBL(1,1) process and their second-order properties

In what follows, we shall focus on the $COBL(1,1)$ subordinate by a Poisson process

Definition 4.2. *The $COBL(1,1)$ process $(X(t))_{t\geq 0}$ delayed by a Poisson process $(N(t))_{t\geq 0}$ is defined by*

$$Y(t) = X(N(t)) \tag{11}$$

that is, the role of time is played by the Poisson process which makes $(Y(t))_{t\geq 0}$ *a Lévy process.*

From the above definition, we can see that there are two sources of randomness: the ground process $(X(t))_{t>0}$ and a time process $(N(t))_{t>0}$. So, it's referred to as a stochastic time change, or 'time deformation'. From the solution (6), it follow that $X(N(t)) = X(0) + \int_0^{N(t)} (\alpha X(s) + \mu)ds + \gamma \int_0^{N(t)} X(s)dw(s), t \geq 0$, then the 1*st* change-of-variable formula yields

$$dY(t) = (\alpha Y(t) + \mu)dN(t) + \gamma Y(t)dw(N(t)), t \geq 0, Y(0) = y_0. \tag{12}$$

Therefore, several authors have considered the process $Y(t) = X\left(\widehat{N}(t)\right)$ where $\widehat{N}(t)$ is the inverse of $N(t)$, *i.e.*;

$$dY(t) = (\alpha Y(t) + \mu)d\widehat{N}(t) + \gamma Y(t)dw\left(\widehat{N}(t)\right), t \geq 0, Y(0) = y_0 \tag{13}$$

(see [19] and the references therein) who gave the connection between the classical Itô *SDE* (4) and their corresponding subordinated *SDE* (12) and (13). The above discussion is summarized in the next lemma

Lemma 4.2. *[Duality of SDEs]. Let N(t) be a Poisson process, then*

a. If $(X(t))_{t\geq 0}$ satisfies the SDE (4) , then $Y(t) = X(N(t))$ satisfies the SDE (12).

b. If $(Y(t))_{t\geq 0}$ satisfies the SDE (13), then $X(t) = Y\left(\widehat{N}(t)\right)$ satisfies the SDE (4).

Proof. See [19].

Now, we are in a position to state the following proposition

Proposition 4.3. *The unique, strong solution to homogeneous SDE* (12) *is explicitely written as*

$$Y(t) = \mathcal{F}(t)\left\{ Y(0) + \mu \int_0^{N(t)} \Phi^{-1}(s)ds \right\}, t \geq 0 \tag{14}$$

where $\mathcal{F}(t) = \exp\{Z(t)\}$ *is the fundamental solutionw with* $Z(t) = \left(\alpha - \frac{1}{2}\gamma^2\right)N(t) + \gamma w(N(t))$.

Proof. It suffices to show that the process $(Y(t))$ given by (14) satisfies *SDE* (12). Set $Y(t) = \mathcal{F}(t)g(t)$ where $g(t) = Y(0) + \mu \int_0^{N(t)} \Phi^{-1}(s)ds$. By the Ito formula and the differential identities we have

$$\begin{aligned} dY(t) &= e^{Z(t)}g(t)dZ(t) + e^{Z(t)}g'(t)dt + \frac{1}{2}\left(e^{Z(t)}g(t)dZ(t) + e^{Z(t)}g'(t)dt\right)' \\ &= Y(t)dZ(t) + \mu dN(t) + \frac{1}{2}Y(t)d[Z,Z] \\ &= Y(t)\left(\left(\alpha - \frac{1}{2}\gamma^2\right)dN(t) + \gamma dw(N(t))\right) + \mu dN(t) + \frac{1}{2}\gamma^2 Y(t)dN(t) \\ &= (\alpha Y(t) + \mu)dN(t) + \gamma Y(t)dw(N(t)). \end{aligned}$$

Thus $Y(t)$ satisfies (12), completing the proof.

Remark 4.4. *If $(X(t))_{t\geq 0}$ is a GMB, then the explicite solution of its subordinated version $(Y(t))_{t\geq 0}$ is $Y(t) = \mathcal{F}(t)Y(0)$, $t \geq 0$ and hence More generally, for any $k \in \mathbb{R}$, we have*

$$E\{Y^k(t)\} = E\{Y^k(0)\} \exp\left\{-\lambda t\left(1 - \exp\left\{\left(\alpha - \frac{\gamma^2}{2}\right)k + k^2\frac{\gamma^2}{2}\right\}\right)\right\}$$

and hence $E\{Y^k(t)\} \to +\infty$ as $t \to \infty$ whenever $\left(\alpha - \frac{\gamma^2}{2}\right)k + \frac{\gamma^2}{2}k^2 > 0$. Additionally,

1. If $\alpha > \frac{1}{2}\gamma^2$, then $\lim\limits_{t\to+\infty} Y(t) = +\infty$,

2. If $\alpha < \frac{1}{2}\gamma^2$, then $\lim\limits_{t\to+\infty} Y(t) = +0$,

3. If $\alpha = \frac{1}{2}\gamma^2$, then asymptotically $(Y(t))_{t\geq 0}$ switches arbitrary between large and small positive values even infinitely.

An extension of Proposition 3.3 for the process $(Y(t))_{t\geq 0}$ is stated in the following proposition

Proposition 4.5. *Let $M_n(t,y) = E\{Y^n(t)|Y(0) = y\}$ the n-th conditional moment of the process $(Y(t))_{t\geq 0}$ defined by* (11) *Then under the condition of proposition 3.3, we have $M_n(t,y) = \sum_{i=0}^{n}\xi_i(n)e^{-\lambda_i^* t}$ where $\lambda_i^* = \lambda(1 - e^{a_i})$ and $\xi_i(n)$ satisfies the recursion* (10).

Proof. From Example 3.2, moments properties of the Poisson process and some manipulation of conditional expectation properties, the results follows.

Example 4.1. *For the $COBL(1,1)$ process delayed by $N(t)$ process defined by* (11) *with fixed initial value, the second-order properties of the process $(Y(t))_{t\geq 0}$ defined by* (11) *are given by*

$$\begin{aligned} E\{M_1(t,y)\} &= -\frac{b_1}{a_1} + P_0(y)\exp\{-\lambda_1^* t\}, \\ t \geq 0 and E\{M_2(t,y)\} &= \frac{b_1 b_2}{a_1 a_2} + P_1(y)e^{-\lambda_1^* t} + P_2(y)e^{-\lambda_2^* t} \end{aligned}$$

where $\lambda_1^* = \lambda(1 - e^{a_1})$, $\lambda_2^* = \lambda(1 - e^{a_2})$. Note that when the initial value is random, the expressions of $E\{Y(t)\}$ and $E\{Y^2(t)\}$ may be obtained by replacing the polynomials $P_0(Y)$, $P_1(Y)$ and $P_2(Y)$ by their expectations. Moreover, it is clear that the first and second moments depends in general on time and on the initial condition, thus the $(Y(t))_{t\geq 0}$ process is not stationary but is asymptotically stationary.

4.3 Distribution

The distribution of the process $(Y(t))_{t\geq 0}$ defined by (11) is given by

$$F_Y(y) = P(X(N(t)) \leq y) = E\{I_{X(N(t))\leq y}\} = E\{E\{I_{X(N(t))\leq y}|N(t)\}\}.$$

Since $X \rightsquigarrow \mathcal{IG}$ with shape $\delta = (\gamma^2 - 2\alpha)/\gamma^2$ and scale $\theta^{-1} = \frac{\gamma^2}{2\mu}$, then each $X(t)$ follows an $I\mathcal{G}(\delta t, \theta)$, that is, has a probability density function (PDF), $f_{X(t)}(x) =$

$\frac{\theta^{t\delta}}{\Gamma(t\delta)}x^{-t\delta-1}\exp\{-\theta/x\}, x>0$ and cumulative distribution function (*CDF*) $F_{X(k)}(x) = \frac{\Gamma\left(t\delta,\frac{\theta}{x}\right)}{\Gamma(t\delta)}$ then the *PDF* and *CDF* functions of $(Y(t))_{t\geq 0}$ are given respectively by

$$f_Y(y) = e^{-\lambda t}I_{\{y=0\}} + e^{-\theta/y}e^{-\lambda t}\frac{1}{y}\sum_{k=1}^{\infty}\left(\left(\frac{\theta}{y}\right)^{\delta}\lambda t\right)^k\frac{1}{k!\Gamma(\delta k)}I_{\{y>0\}}$$

$$F_Y(y) = H(y)e^{-\lambda t} + e^{-\lambda t}\sum_{k=1}^{\infty}\Gamma(\delta k, \theta/y)\frac{(\lambda t)^k}{k!\Gamma(\delta k)}.$$

where $H(.)$ is the Heaviside step function, therefore, the probability law of $(Y(t))_{t\geq 0}$ has atom $e^{-\lambda t}$ at zero, that is, has a discrete part $P(Y(t)=0) = e^{-\lambda t}$.

Remark 4.6. *An equivalent expression of the above PDF and CDF functions may be given by the following Poisson mixture:* $f_Y(y) = \sum_{k=0}^{\infty} f_{X(k)}(y)P(N(t)=k)$, $F_Y(y) = \sum_{k=0}^{\infty} F_{X(k)}(y)P(N(t)=k)$. *These PDF and CDF function are the same as for* $Z(t) = \sum_{n=1}^{N(t)}\xi_n$ *where* $(\xi_n)_{n\geq 1}$ *is a sequence of i.i.d. random variables independent of* $N(t)$. *Note that when* $(X(t))_{t\geq 0}$ *and* $(N(t))_{t\geq 0}$ *are independent processes and the relevant moments exist, then* $E\{Y(t)\} = t\mu_X\mu_N$ *and* $Var\{Y(t)\} = t\left(\sigma_N^2\mu_X^2 - \sigma_X^2\mu_N\right)$ *where* $\mu_N = E\{N(1)\}$, $\mu_X = E\{X(1)\}$, $\sigma_N^2 = Var(N(1))$ *and* $\sigma_X^2 = Var(X(1))$.

The $\mathcal{IG}$ distribution belongs to the exponential family of distribution with respect to $\theta = (\delta+1,\theta)'$. Indeed, $f_{\mathcal{IG}}(x) = \frac{\theta^{\delta}}{\Gamma(\delta)}x^{-\delta-1}\exp\{-\theta/x\} = \exp\{-\underline{\theta}'T(x) + A(\theta)\}$ where $T(x) = \left(\log(x), \frac{1}{x}\right)'$ and $A(\theta) = \delta\log(\theta) - \log(\Gamma(\delta))$. The function $A(.)$ is known as the cumulant function its first and second derivative provide the mean and the variance of $T(X)$. So $f_{Y(t)}(y)$ may be rewritten as

$$f_{Y(t)}(y) = e^{-\lambda t}I_{\{y=0\}} + e^{-\lambda t}\sum_{k=1}^{\infty}\exp\{-\underline{\theta}'(k)T(x) + A(\theta(k))\}P(N(t)=k)I_{\{y>0\}}$$

in which the vector $\underline{\theta}(k)$ is obtained by replacing the parameter β in $\underline{\theta}$ by $k\beta$. So, the distribution of $(Y(t))_{t\geq 0}$ may be regarded (asymptotically) as the distribution of *GIG* subordinated by the Poisson process. Regardless of the form of the expected value of the function $h(Y)$ is expressed as $E\{h(Y)\} = \int_{\Theta} E_{X|N=k}\{h(Y)\}g_N(k)d\nu(k)$ where $E_{X|N}\{.\}$ is taken with respect to the conditional distribution of X. In particular, $E\{Y\} = E\{E_{X|N}(X)\}$ and $Var\{Y\} = Var\{E_{X|N}(X)\} + E\{Var_{X|N}(X)\}$. Moreover

$$E\{Y^n\} = e^{-\lambda t}\sum_{k=0}^{\infty}\left\{\int_0^{\infty} y^{n-\delta k-1}e^{-\theta/y}dy\right\}\theta^{\delta k}(\lambda t)^k\frac{1}{k!\Gamma(k\delta)}$$

$$= \theta^n e^{-\lambda t}\sum_{k=0}^{\infty}\frac{\Gamma(k\delta-n)}{\Gamma(k\delta)}\frac{(\lambda t)^k}{k!}$$

$$= \theta^n e^{-\lambda t}{}_1\Psi_1(\delta, -n, \delta, 0, \lambda t)$$

where ${}_1\Psi_1(\rho, a, \rho, b, x) = \sum_{k=0}^{\infty} \frac{\Gamma(\rho k + a)}{\Gamma(\rho k + b)} \frac{x^k}{k!}$ is the confluent hypergeometric function that plays an important role in mixing theory. For certain values of the parameter ρ and for $n > 0$, it is possible to give representations of ${}_1\Psi_1(\rho, a; \rho, 0; x)$ in terms of well-known special functions. In general, the exact expression of ${}_1\Psi_1(\rho, a, \rho, b, x)$ is very difficult to express it, so in literature, the solution is given for certain specific case, (interested readers are advised to see [20] and the references therein). The solution of ${}_1\Psi_1(\rho, a, \rho, b, x)$ is a vast subject and we will not develop it further here.

5. Estimation issues

In this section, we propose the moment's method (*MM*) for estimating the unknown parameters, α, μ and γ gathered in vector $\underline{\theta}$ involved in $COBL(1,1)$ and in its distribution $\mathcal{IG}$. The estimates parameters according to *MM* are obtained from two processes $(X(t))_{t\geq 0}$ and $(Y(t))_{t\geq 0}$. Moreover, we concentrated on the weak and/or asymptotically stationary case and we assume that the parameter λ in the Poisson process is known. The first and second moments of the asymptotically stationary process $(Y(t))$ as defined in Example 4.1 are $\mu_1 = -\frac{b_1}{a_1} = -\frac{\mu}{\alpha}$, and $\mu_2 = \frac{b_1 b_2}{a_1 a_2} = \frac{-\mu^2}{\alpha(2\alpha+\gamma^2)}$. Additionally, from proposition 3.1, we have asymptotically $\rho(1) = e^{\alpha}$. So the following formulas for the parameters can be derived $\alpha = \log \rho(1)$, $\mu = -\alpha\mu_1$, and $\gamma^2 = -\frac{(\mu^2 + 2\alpha\mu_2)}{\mu_2}$. These relationships can be used for estimating $\underline{\theta}$ by *MM*, more precisely the estimators are given by

$$\widehat{\alpha} = \widehat{\log\rho(1)}, \widehat{\mu} = -\widehat{\alpha}\widehat{\mu}_1 \text{and} \widehat{\gamma^2} = -\widehat{\alpha}\frac{(\widehat{\mu}_1^2 + 2\widehat{\mu}_2)}{\widehat{\mu}_2}$$

where $\widehat{\mu}_1$, $\widehat{\mu}_2$, and $\widehat{\log\rho(1)}$ are respectively the empirical first, second-order moment, and the empirical logarithm of autocorrelation. Their consistency and asymptotic normality are given the following proposition

Proposition 5.1. *Under the Assumption 1,we have*

1. $\left(\widehat{\underline{\theta}}_n\right)$ *converges in probability to* $\underline{\theta}_0$

2. $\sqrt{n}\left(\widehat{\underline{\theta}}_n - \underline{\theta}_0\right) \rightsquigarrow N(\underline{O}, \Sigma(\underline{\theta}_0))$ *where* $\Sigma(\underline{\theta}_0)$ *is* 3×3 *asymptotic covariance matrix.*

Proof. The proof follows essentially the same arguments as in Bibi and Merahi [21].

5.1 Some simulation results

In order to check the effectiveness of the described estimation procedure, we simulated 500 trajectories of length $n \in \{1000, 2000\}$ with parameters $\underline{\theta}$ shown at the bottom of each table below. The vector $\underline{\theta}$ is chosen to satisfy the second-order stationarity and the existence of moments up to fourth-order. For the purpose of illustration, the vector of parameters $\underline{\theta}$ is estimated with $(X(t))_{t\geq 0}$ noted $\widehat{\underline{\theta}_m^{(X)}}$ and

	n = 1000		n = 2000		n = 1000		n = 2000	
$\underline{\theta}$	Mean		Mean		Mean		Mean	
$\widehat{\alpha}^{(X)} = \left(\widehat{\alpha_m^{(X)}}, \widehat{\alpha_g^{(X)}}\right)$	−1.5461 (0.0371)	−1.6665 (0.0359)	−1.6608 (0.0365)	−1.5864 (0.0344)	−0.6015 (0.0502)	−0.4906 (0.0429)	−0.5702 (0.0175)	−0.5501 (0.0162)
$\widehat{\mu}^{(X)} = \left(\widehat{\mu_m^{(X)}}, \widehat{\mu_g^{(X)}}\right)$	0.2550 (0.0702)	0.2763 (0.0681)	0.2753 (0.0601)	02639 (0.0672)	1.0394 (0.0612)	1.0291 (0.0601)	1.0311 (0.0165)	1.1151 (0.0152)
$\widehat{\gamma}^{(X)} = \left(\widehat{\gamma_m^{(X)}}, \widehat{\gamma_g^{(X)}}\right)$	0.7135 (0.0581)	0.7428 (0.0770)	0.7335 (0.0475)	0.74994 (0.0692)	0.5066 (0.0571)	0.4989 (0.0459)	0.5011 (0.0122)	0.5002 (0.0201)
$\widehat{\alpha}^{(Y)} = \left(\widehat{\alpha_m^{(Y)}}, \widehat{\alpha_g^{(Y)}}\right)$	−1.4122 (0.0710)	−1.4286 (0.0654)	−1.4466 (0.0705)	−1.4552 (0.0612)	−0.6080 (0.0683)	−0.6086 (0.0612)	−0.5287 (0.0475)	−0.5106 (0.0452)
$\widehat{\mu}^{(Y)} = \left(\widehat{\mu_m^{(Y)}}, \widehat{\mu_g^{(Y)}}\right)$	0.2261 (0.0725)	0.2311 (0.0664)	0.2387 (0.0675)	0.2413 (0.0612)	1.1120 (0.0691)	1.1013 (0.0682)	1.0195 (0.0445)	1.0200 (0.0431)
$\widehat{\gamma}^{(Y)} = \left(\widehat{\gamma_m^{(Y)}}, \widehat{\gamma_g^{(Y)}}\right)$	0.6914 (0.0681)	0.7236 (0.0617)	0.7091 (0.0675)	0.7381 (0.0601)	0.4280 (0.0691)	0.4916 (0.0652)	0.4399 (0.0462)	0.4966 (0.0422)
	Design(1):$\underline{\theta} = (-1.5, 0.25, 0.75)'$				Design(2): $\underline{\theta} = (-0.5, 1.0, 0.5)'$			

Table 1.
The MM *and* GMM *estimation of the processes* X(t) *and* Y(t).

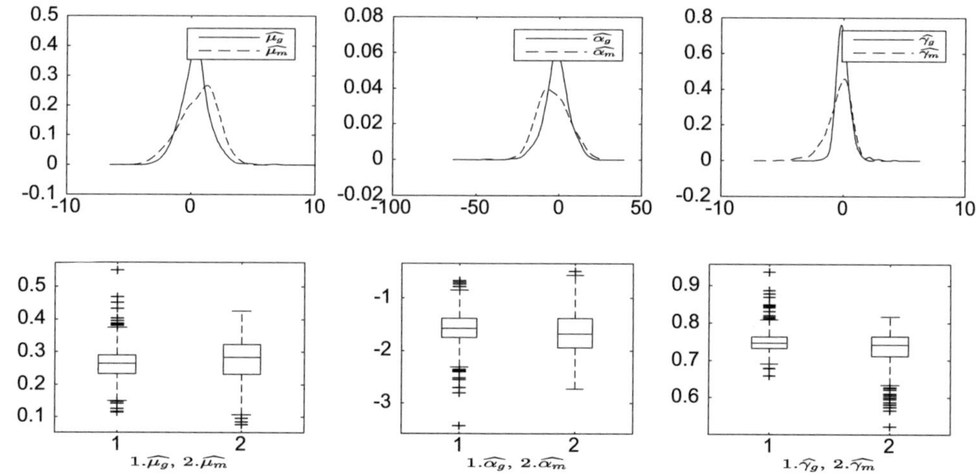

Figure 1.
*Top panels: The overlay of asymptotic kernel of $\sqrt{n}(\hat{\theta}_m(i) - \theta(i)$and $\sqrt{n}(\hat{\theta}_g(i) - \theta(i)$ based on X(t). Bottom panels: Are the corresponding boxplots summary of $\hat{\theta}_m(i)$and $\hat{\theta}_g(i), i = 1, 2$ and 3 according to Design (1) illustrated in **Table 1**.*

	n = 1000	n = 2000
	$\widehat{\text{Scale}}$, $\widehat{\text{Schape}}$	$\widehat{\text{Scale}}$, $\widehat{\text{Schape}}$
MM	(1.002, 7.073)	(1.023, 7.173)
GMM	(1.001, 7.040)	(0.938, 6.641)
Design(1): Scale: 0:8889 and Schape: 6:3333		

Table 2.
The estimation of the distribution of X(t).

compared with its delayed $(Y(t))_{t\geq 0}$ process noted $\widehat{\underline{\theta}_m^{(Y)}}$.As a parameter of configuration we estimate $\underline{\theta}$ by the generalized method of moment (GMM) noted $\widehat{\underline{\theta}_g^{(X)}}$ and $\widehat{\underline{\theta}_g^{(Y)}}$. In the Tables below, the column "Mean" correspond to the average of the parameters estimates over the 500 simulations. In order to show the performance of the estimators, we have reported in each table the root means squared error ($RMSE$) (results between brackets). The results of estimating corresponding to the process $(X(t))$ and $(Y(t))$ are summarized in **Table 1.**

The plots of the asymptotic density of each component of $\widehat{\underline{\theta}}$ according to MM and GMM methods based on process $(X(t))$ (resp. on process $(Y(t))$) are summarized in the **Figure 1** (resp. in **Figure 3**)

Additionally, the estimates of scale and shape parameters of $\mathcal{IG}$ distribution are reported in **Table 2.**

The plot of estimate $\mathcal{IG}$ distribution of the process $X(t)$ is shown in **Figure 2.**

5.2 Comments

Now a few comments can be made

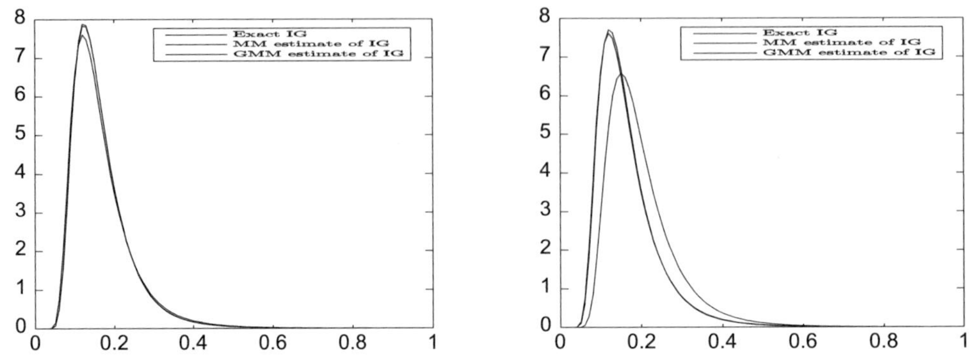

Figure 2.
The left plot is the overlay of exact, MM estimate and GMM estimate associated to design (1) of IG *distribution of the process* X(t) *with* n = *1000. The right plot is similar to the left with* n = *2000.*

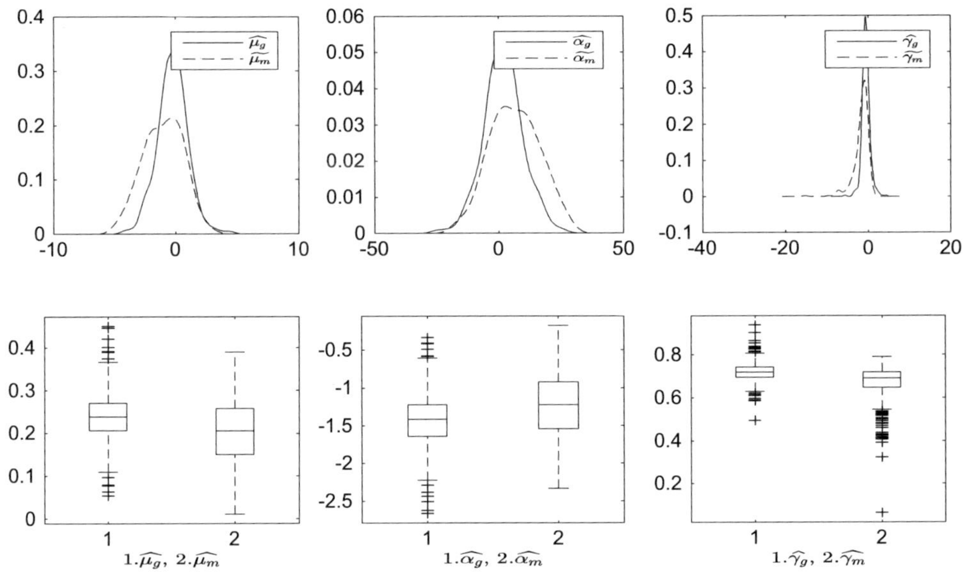

Figure 3.
Top panels: The overlay of asymptotic kernel of $\sqrt{n}(\hat{\theta}_m(i) - \theta(i)$ *and* $\sqrt{n}(\hat{\theta}_g(i) - \theta(i)$ *based on* Y(t)*. Bottom panels: Are the corresponding boxplots summary of* $\hat{\theta}_m(i)$*and* $\hat{\theta}_g(i), i = 1, 2$ *and 3 according to Design (1) illustrated in* ***Table 1****.*

A. By inspecting **Table 1**

1. it is clear that the results of *MM* and *GMM* methods are reasonably closed to the true values and their *RMSE* decreases when the sample size increases.

2. The above observations may be seen by regarding the plots of asymptotic distributions of their kernels estimates displayed in **Figure 2** showing the moderate-fat tails (positive kurtosis or leptokurtic) of such a kernels and the asymptotic accuracy of *MM* and *GMM* estimates.

3. Additionally, it can be seen from the boxplots displayed in **Figure 2** that the methods *MM* displays more outliers than *GMM*. This is not surprising due

to the robustness properties of *GMM* and hence its capability to detect the outliers in nonlinear models.

4. It can be observed that the *RMSE* associated with *GMM* is more less than of that associated to *MM*.

5. It can be said that the estimation of scale parameters is more accurate for the smaller values of those parameters whereas the estimation of shape parameters is more accurate for the larger values of those parameters.

B. By inspecting **Table 2**

1. The performances of *GMM* and *MM* are according to their order, and are close to each other.

2. It seems that it is very difficult to distinguish between the results reported in **Table 1** and **Table 2** and the plots of the asymptotic kernels showed in **Figures 1** and **3**. This is due to asymptotic stationary which led to the same parameters involved in both *SDE* (4) and (12).

3. For $n = 1000$ and/or $n = 2000$, it is observed that *GMM* works the best from *MM* for both designs of the two parameters α and μ.

6. Concluding remarks and future research direction

The stochastic subordination model proposed in this paper is practically and theoretically appealing for the modeling of several phenomena already pointed out in Section 1. Such models are rich enough to model among others, the observed non-normal returns, significant autocorrelation of squared returns. In this paper, we have proposed a theoretical model that not only takes under consideration such specific property but also exhibits short-range dependence and can be used for data with visible jumps. This model is based on the stable $COBL(1,1)$ process delayed the Poisson subordinator. The proposed model is non-linear and non-normal, involves three additional parameters which may easily and quickly be estimated under the asymptotically stationarity assumption with a moments method (*MM)* and compared with a generalized method of moments (*GMM*). Clearly, the analyzed process is complex and the estimation is challenging. A significant advantage of the stochastic subordination model is that it inherits some properties of the process to be subordinated and hence the stationary and nonstationary process can be obtained through the subordination approach. These issues are of importance to theoreticians and practitioners alike and will be the subject of further papers. Further research is required to investigate the asymptotic theory of estimator under more matching conditions. The model presented in this paper may be slightly modified by replacing the Poisson process by other processes subject to some appropriate condition.

Classification

2010 AMS Math. Subject Classification: Primary 40A05, 40A25; Secondary 45G05.

Author details

Abdelouahab Bibi
Department of Mathematics, Larbi Ben M'hidi University, O.E.B., Algeria

*Address all correspondence to: abd.bibi@gmail.com

References

[1] Mohler RR. Nonlinear Time Series and Signal Processing. Berlin: Springer Verlag; 1988

[2] Aït-Sahalia Y. Estimating continuous-time models with discretely sampled data. In: Blundell R, Persson R, Newey W, editors. Econometrics Theory and Applications. Cambridge: Cambridge University Press; 2007. pp. 261-327

[3] Le Breton A, Musiela M. A study of one-dimensional bilinear differential model for stochastic processes. Probability and Mathematical Statistics. 1984;**4**:91-107

[4] Bibi A, Merahi F. A note on L_2-Structure of continuous-time bilinear processes with time-varying coefficients. International Journal of Statististics and Probability. 2015;**4**:150-160

[5] Lee MLT, Whitmore GA. Stochastic processes directed by randomized time. Journal of Applied Probability. 1993;**30**: 302-314

[6] Clark PK. A subordinated stochastic process model with finite variance for speculative prices. Econometrica. 1973; **41**:135-155

[7] German H, Ane T. Stochastic subordination. RISK. 1996;**9**:146-149

[8] Aït-Sahalia Y, Jacod J. High-Frequency Financial Econometrics. New Jersey: Princeton University Press; 2014

[9] Klüppelberg C, Lindner A, Maller R. A continuous time GARCH process driven by a Lévy process: Stationarity and second order behaviour. Journal of Applied Probability. 2004;**41**:601-622

[10] Brockwell PJ. Continuous-time ARMA processes. In: Shanbhag DN, Rao CR, editors. Handbook of Statistics. Amsterdam: North Holland; 2001. pp. 249-276

[11] Ksendal B. Stochastic Differential Equations: An Introduction with Applications. New York: Springer-Verlag; 2000

[12] Iglói E, Terdik G. Bilinear stochastic systems with fractional Brownian motion input. The Annals of Applied Probability. 1999;**9**:46-77

[13] Leon JA, Perez-Abreu V. Strong solutions of stochastic bilinear equations with anticipating drift in the first Wiener chaos. In: Cambanis S, Ghosh JK, Karandikar R, Sen PK, editors. Stochastic Processes: A Festschrift in Honor of Gopinath Kallianpur. Berlin: Springer-Verlag; 1993. pp. 235-243

[14] Arnold L. Stochastic Differential Equations, Theory and Applications. New York: John Wiley; 1974

[15] Bishwal J, Jaya PN. Parameter Estimation in Stochastic Differential Equations. Berlin: Spring-Verlag; 2008

[16] Has'minskii RZ. Stochastic Stability of Differential Equations. Sijthoff \& Noordh. Berlin, Heidelberg: Springer-Verlag; 1980

[17] Bibi A, Merahi F. Yule-Walker type estimator of first-order time-varying periodic bilinear differential model for stochastic processes. Communication Statististics: Theory and methods. 2020; **49**:4046-4072

[18] Ikeda N, Watanabe S. Stochastic Differential Equations and Diffusion Processes. Tokyo: North-Holland/ Kodansha Ltd.; 1981

[19] Kobayashi K. Stochastic calculus for a time-changed semimartingale and the associated stochastic differential equations. Journal of Theoretical Probability. 2011;**24**:789-820

[20] Paris RB, Vinogradov V. Asymptotic and structural properties of special cases of the Wright function arising in probability theory. Lithuanian Math. 2016;**56**:377-409

[21] Bibi A, Merahi F. Moment method estimation of first-order *continuous*-time bilinear processes. Communications in Statistics: Simulation and Computation. 2019;**48**:1070-1087